B. F. SKINNER
A LIFE

OTHER BOOKS BY DANIEL W. BJORK

The Victorian Flight:
Russell Conwell and the Crisis of American Individualism (1978)

The Compromised Scientist:
William James in the Development of American Psychology (1983)

William James: The Center of His Vision (1988)

B. F. Skinner

A Life

DANIEL W. BJORK

BasicBooks
A Division of HarperCollins*Publishers*

Designed by Ellen Levine

Library of Congress Cataloging-in-Publication Data
Bjork, Daniel W.
B. F. Skinner: a life / by Daniel W. Bjork.
p. cm.
Includes bibliographical references and index.
ISBN 0-465-00611-6
1. Skinner, B. F. (Burrhus Frederic), 1904–1990.
2. Psychologists—United States—Biography. 3. Behaviorism (Psychology)—United States—History. I. Title.
BF109.S55B46 1993
150.19′434′092—dc20
[B] 92-54522
CIP

93 94 95 96 CC/HC 9 8 7 6 5 4 3 2 1

To the memory of

Carla Annette Brown
and
Randall Elvin Brown

who wanted a better world, helped others find one,
and will not be forgotten by
those who loved them.

I am a humanist in the sense that nothing human is alien to me.
—Interview with B. F. Skinner, March 9, 1990

CONTENTS

PREFACE

B.F. Skinner's life spanned most of the twentieth century and has served as a lightning rod for American opinion. Some view him as a reductive, mechanistic behavioral scientist who denied the existence of a creative, purposeful mind or an inner person free to choose and accept responsibility for one's actions. He wrongly equated the behavior of rats and pigeons with humans, they say, maintaining that despite the obvious higher mental capacity of the latter, all organisms could be controlled or manipulated through a psychology of behavior called positive reinforcement. By belittling superior mentality and disavowing free choice, Skinner degraded what was most human in humanity. His behavioral engineering brainwashed individuals into enjoying being conditioned, thereby eliminating not only the distinction between people and animals but personal freedom and dignity as well. To his most fervent opponents, Skinner was the Darth Vader of American psychology, perhaps even the Hitler of late-twentieth-century science itself—a man whose science of conditioning threatened the dearest humanistic traditions, indeed, those that made life most worth living.

For others, however, Skinner was the brilliant originator of radical behaviorism, a science that yielded the most controllable and hence most predictable experimental results in the history of psychology. Moreover,

Skinner took his science out of the laboratory and into the world, where it could help people. Far from being reductive and mechanistic, Skinner was expansive and innovative. His science has done much more to help people than any of the so-called humanist psychologies or philosophies. Not only behaviorists but mental health professionals, teachers, and businesspeople have used reinforcement techniques to improve the quality of life for the mentally retarded and the addicted, to upgrade classroom performance, and to improve the morale and productivity of workers. Techniques of positive reinforcement and cultural design could, if applied over time and on a grand scale, save the world from the catastrophes of urban decay, ecological ruin, and uncontrolled population growth. Far from being an evil, antihumanist scientist, Skinner was the greatest humanistic scientist of our time, a scientific savior whose legacy to the human species—a practical behaviorist technology—is currently helping thousands and has the potential to produce a better life for us all.

In writing this biography, I have tried not to take sides; I will not seek to prove Skinner's detractors or his supporters wrong. While attempting to present an evenhanded appraisal of his life and work, I have kept the central focus on the relationship between Skinner and the American tradition. Like him or not, Skinner had something worth saying about the possibilities and limits of American life, because he experienced America in a common and yet, as it turned out, special way. He grew up in a small town in the early twentieth century and shared the innocence and optimism of that milieu. But he was also an alienated and cynical intellectual in the 1920s, parting company with mainstream culture, especially its boosterism and business orientation. Another persona, however, came to dominate his career as a behavioral scientist. Skinner became the American-as-inventor, a man fascinated with devising gadgets, an inventor whose optimism and mechanical cleverness allowed him to find and develop not only a new science but also a novel American technology of social invention with which he hoped to design a better world.

It is time to situate Skinner among the galaxy of American intellectuals and inventors who represent different facets of the national intellectual and cultural tradition—stars such as Jonathan Edwards, Benjamin Franklin, Ralph Waldo Emerson, Henry David Thoreau, William James, Thomas Edison, and John Dewey. Skinner's star cast an unusual light. Whether or not he offered a world of promise or fright, he was an American original, adding a fresh twist to the American scientific, intellectual, and social heritage. As a new century approaches, it is particularly appropriate to reassess tradition. Skinner allows a splendid biographical opportunity to do so.

RESEARCH FOR THIS BOOK was facilitated by stipends from the National Endowment of the Humanities and the American Philosophical Society, by a grant from the University of Detroit Mercy, by a reduced teaching load at St. Mary's University, and by advances from the publisher, Basic Books.

The staffs of the Harvard Archives, Pusey Library, Harvard University; the Schlesinger Library on the History of Women in America, Radcliffe College; the Walters Library, University of Minnesota; the Hamilton College Library, Clinton, New York; and the Susquehanna Free Library, Susquehanna, Pennsylvania, were kindly, efficient, and indispensable. Special appreciation goes to Frank Lorenz, whose knowledge of Skinner-related materials in the Hamilton College Library greatly aided my understanding of the undergraduate Fred Skinner. Skinner's own willingness to be interviewed and to give access to personal notes and other materials in his basement archive was generous and invaluable. Other family members were also cooperative and helpful. Eve Skinner's interviews with my wife, Rhonda Bjork, yielded a richer understanding of her husband and the Skinner family, as did my own discussions with Julie and Ernest Vargas, Skinner's older daughter and her husband, and with Deborah Buzan, the younger daughter. Conversation with Fred S. Keller, who shared a science and friendship with Skinner for over fifty years, was invaluable in gaining biographical information and perspective.

I am indebted to several individuals at Indiana University (Skinner chaired the department of psychology there in the late 1940s), including Eliot Hearst, Douglas Ellson, George Heise, and James Dinsmoor, who in the early stages of research agreed to talk about Skinner, whom each had known. Hearst, a historian of psychology, encouraged me to write to Skinner a second time when my first correspondence went unanswered because Skinner had taken a fall and was recuperating in the hospital.

Thanks to my old friend Terry Wallenbrock, who in the early 1980s first suggested the cultural and intellectual importance of doing a biography on B. F. Skinner. The encouragement of historians Robert Ferrell and Paul Varg helped keep the project a first priority when career considerations might have derailed it. Conversations and/or correspondence with Max Brill, Stephen Coleman, Richard Herrnstein, Jay Moore, Edward Morris, Laurence D. Smith, Nils Wiklander, and William R. Woodward helped define the intellectual context of behavioral analysis as well as offer insights about Skinner's personal characteristics and professional career. Talented critics with more pleasant and profitable things to do read parts or all of the first draft, including James A. Dinsmoor, Daniel Fallon, Fred S. Keller, David W. Levy, Edward Morris, Clifford Scott, Laurence D. Smith, Ernest Vargas, Julie S. Vargas, William R. Woodward, and Jacqueline Zeff. It is a much better book for their efforts.

Two individuals had a particularly salutary effect. The historian of science Laurence D. Smith sent a thoughtful commentary on a 1989 essay that attempted to place Skinner in the American intellectual tradition and assess him as a social inventor—two major themes in this book. Without Smith's enthusiasm and critical comments, the biography may well have lost focus or never been written. James A. Dinsmoor, suffering from a back injury, read every word from a behaviorist perspective but with sensitivity to the lay point of view, and offered especially useful and insightful suggestions for the final revision.

Appreciation also goes to those whose technical skills, time, care, and professional advice helped transform the manuscript into a book: to Elizabeth Pangrazzi for transcribing Skinner interviews; to Elizabeth Szalay for interlibrary loan assistance; to Frank Lorenz, Scott Schrader, and Julie S. Vargas for help with photographs; to Basic Books editor Susan Arellano for approaching me about doing a Skinner biography, and senior editor Jo Ann Miller for expert criticism on first-draft chapters; to editorial assistant Melanie Kirschner for assisting with small and not so small details; to superb copyeditor Linda Carbone; to skilled project editor Jane Judge and indexer Steve Csipke; and to Basic's president, Martin Kessler, for reading chapters and supporting the completion of the manuscript while it was between editors.

And thanks most of all, Rhonda, for persevering on the project through the loss of a sister and a brother. Your interviewing, research, and help with revision were more than indispensable; you were unforgettably courageous and supportive. It is your book too, and better in all ways for you.

B. F. SKINNER
A LIFE

1

Inventive Beginnings

Making do . . . that has always been a favorite theme of mine. To make the most of what you have.

—B. F. Skinner, Basement Archives, 1971

Down a narrow basement staircase in a one-story ranch-style house in the Larchmont area of Cambridge, Massachusetts, about two miles from Harvard Yard, is a rectangular-shaped study. Near the study door there are comfortable armchairs, one equipped with movable metal arms fitted with a reading lens. Toward the far end of the study, facing each other on opposite walls, are a long wooden writing desk and a bright yellow sleeping cubicle, complete with stereo system, storage compartment for musical tapes—especially Wagner—and a timer which, with circadianlike rhythm, rang at five o'clock every morning for over twenty years to bring B. F. Skinner to his writing desk, like a monk to his matins. For two hours every morning, until the timer rang again at seven, one of America's most controversial intellectuals worked on the papers, articles, and books that would define and defend a science he called the experimental analysis of behavior.

Here in his study he had arranged a boxlike environment that enabled him to manage the intellectual behavior of his own organism practically up to his final moments. Indeed, three days before his death on August 18, 1990, Skinner was at his desk, answering correspondence and thinking about the reactions to a speech he had just delivered to the American Psychological Association in Boston in which he had compared the failure of cognitive psychologists to accept the science of behavioral analysis to the failure of nineteenth-century creationists to accept Darwinian natural selection.

Skinner's study with its sleeping cubicle was the last boxlike world he had invented, places he had first fashioned out of cardboard to read in as a boy. But it was as a scientist that the box motif became identified with Skinner, beginning with the operant chamber, or "Skinner box," in which he could observe and record the behavior of rats, much as other scientists watched microbes move through a microscope. He would continue to fashion a world of boxes as social inventions to improve human living: the "baby tender," or "aircrib," which allowed an infant unrestrained freedom of movement in a thermostatically controlled space; a small fictional community called Walden Two in which behavioral engineering created a place where people no longer needed the open sprawl of large cities; and the teaching machine, a mechanical device shaped like a box that was "programmed" to use the behavioral technique of positive reinforcement to facilitate student learning. Skinner's life as a scientist, social inventor, and intellectual was inexorably tied to a world of boxes—environments that controlled or selected the behaviors of which they were a function.

In a voice barely above a whisper, broken by coughing, Skinner insisted days before his death that most of the major turning points of his life and the discovery of his science had been sheer accident. He had been exceedingly fortunate in his personal life, yet he developed a behavioral science to change environmental contingencies to achieve self-control. His world of boxes produced the chance to achieve remarkable control, control of the behavior of rats, of pigeons, of people, and of himself. The life of B. F. Skinner is a study in the juxtaposition of chance and control, of the accidental and the determined, the story of a man who respected and even courted chance but who sought with sustained diligence, even obsessiveness, to shape a better world scientifically.

BURRHUS FREDERIC SKINNER was born in 1904 and spent his first eighteen years in the small northeast Pennsylvania town of Susquehanna, a few miles south of the New York State border. Susquehanna, named after the river that meanders in great loops through eastern Pennsylvania and southern New York, maintained a population of around two thousand and was overshadowed by two modest regional centers: Binghamton, New York, fifteen miles north; and Scranton, Pennsylvania, thirty miles south. Because its residents lived on the abruptly rising hills that flanked the narrow Susquehanna river valley running parallel to the Erie Railroad tracks, Susquehanna was known as "the city of stairs." Several blocks from the railroad is that ubiquitous American roadway, Main Street, lined with perhaps a dozen one- and two-story brick and wooden businesses. Off Main Street is an elongated hill, traversed by Grand Street.

Young Fred lived with his parents, William and Grace, and his younger brother, Edward, called "Ebbie," at 433 Grand. The square two-story wooden house was cold in the winter, heated only by a coal furnace that forced hot air up through a large grid in the floor between the living and dining rooms. Fred remembered "standing with my mother and brother on the grid as the first warm air came up in the morning—all of us shivering."[1] The living room was a social but rather formal place where his parents sat in the evenings when he and Ebbie came to say goodnight and where his father sat when Fred "confessed the shortage in my accounts" after he had taken petty cash without explanation. The parlor was more relaxed, the place where the children played, where the Christmas tree was displayed, and where the piano and Victrola stood. In the kitchen, milkshakes and fondants—thick, creamy syrup candies—were made.[2] Next door was the Grand Street cemetery, where neighborhood children played among the stone testimonials to Susquehanna's departed. Fred and Ebbie were cautioned by their mother never to step on a grave. Behind the house was Billy Main's blacksmith shop yard, where abandoned automobile chassis rested.[3]

By his own description, Fred lived in "chaotic conditions under which children learn[ed] to explore, to organize, to select, to construct without a plan," a sort of anarchist out-of-doors environment where children and animals roamed neighbors' yards and there was no such thing as trespassing:[4] "Our yard was a mess, the town was a mess, the surrounding countryside was largely primeval or on its way back to that condition." The Skinners' backyard was strewn with debris, the garden was overgrown, and the

> driveway, ingeniously leading *through* the garage and back to the street on the other edge of the lot, was never well kept. We used it for measured footraces.

> The corner back of the garage was a jumble of currant bushes and rhubarb. We dug deep holes in the beautifully sandy soil and piled up mountains of excavated sand. We built shacks—at one time from seasoned, red-painted boards acquired when a new fence went up alongside the cemetery. Heavy oak planks, oil-soaked, from the floor of the railroad shops, made slides (rather splintery) and merry-go-rounds.[5]

This disarray presented an enterprising, clever boy with myriad opportunities to make things. Yet it was not the disorder alone that engaged Fred's abilities. It was also the opportunity of "making do" with materials at hand. In those days there were few prepackaged toys or amusements. One did not buy adventure so much as make it. Fred read Jules Verne's *The Mysterious Island* as a novel not so much about a scientific future as about mundane tinkering and problem solving. "Like *Robinson Crusoe* and the *Swiss Family Robinson,*" he later wrote, "it was concerned with making do. And that has always been a favorite theme of mine."[6] Invention was improvised in a disorderly, messy setting. During his boyhood Fred would spend untold hours exploring, puttering, tinkering, and building; his world not only allowed but encouraged him to "make do" in these ways. Young Skinner seldom complained of boredom, that constant refrain of late-twentieth-century American children.

From the beginning Fred viewed invention as he would science: a matter of improvisation and accidental discovery rather than a premeditated process of ordering the environment.[7] When the desk in his study was clean, he remarked, he had difficulty discovering what he wanted to say.

SUSQUEHANNA, incorporated as Susquehanna Depot in 1853, developed because the Erie Railroad had come to the area in the 1840s. Railroad expansion in the eastern United States was the last phase of Jacksonian America's transportation revolution, the creation of a national market economy linking canals, turnpikes, and railroads. By 1851 the Erie was the longest railroad in the world owned by one company, extending from New York to the Great Lakes. It was called the Lion of the Railway.[8] Susquehanna grew notably during the Civil War when the Erie built a roundhouse and locomotive turntable there. Like hundreds of American towns, Susquehanna existed to serve the railroad. Many citizens, including large numbers of Irish Americans, worked in the Erie shops. The town awoke, reported for work, and even answered fire emergencies to the sound of the Erie whistle. The railroad and its repair shops fascinated young Fred Skinner, who marveled at the machinery and stared in awe at the enormous Matt Shay, the largest steam engine locomotive in the world, once stationed in Susquehanna.

Of course, a railroad town had its disadvantages, too. Susquehanna had its share of grime, dirt, and pollution. One of Fred's contemporaries wondered, "who could keep a railroad town clean? Without the cinders there would have been no living."[9] Another resident recalled that the town's furnaces burned soft coal, which gathered on windowsills and "hung in the air ruining laundry hung outside." Wooden houses and buildings frequently caught fire, adding further pollutants to the air.[10]

American railroad towns liked to be associated with the progress of industry. The pastor of the Susquehanna Presbyterian Church noted that "many Chambers of Commerce boast 'the smallest big city in America.' . . . It fit Susquehanna—its heavy industry, its incomparable rail connections, and its daily newspaper—all the makings of a metropolis, but by its location doomed to the fate of Tom Thumb. . . . Susquehanna was both cosmopolitan (the American melting pot) and provincial."[11] Although retaining some of the traditional village conviviality, towns like Susquehanna were oriented toward serving the new industrial cities. Indeed, Susquehanna suffered some of the same dislocations that large cities did, experiencing the national railroad strike of 1877, a boilermakers' lockout in 1901, a general machinists' strike in 1907, and an all-crafts strike in 1922. Tragedy hit its citizens in the form of a smallpox epidemic, accidental railroad deaths, and drownings. "Fear of encountering highwaymen" was another problem. The *Susquehanna Transcript* reported in 1905 that three recent holdups had made residents reluctant to walk the streets alone at night. Stealing from local businesses, whether from need (as when poor Italian families took coal from a railroad car) or pure thievery, was a persistent concern. So was vice, which resulted in periodic closing of the gambling houses and poolrooms and raids on "bawdy houses."[12]

Among those who arrived in Susquehanna Depot after the Civil War were Fred's paternal and maternal grandparents. James Skinner, born in Devonshire, England, came to America with his two half-brothers in the early 1870s. After living for a time in New York City, he moved to

Pennsylvania and settled in Starrucca, a village near Susquehanna. There he met and married Josephine Penn, the daughter of an impoverished farmer who barely eked out a living for himself, his wife, and his twelve children. Josephine claimed to be a descendent of William Penn, founder of Pennsylvania, but this grand lineage was never proved. After a sojourn to Amesbury, Massachusetts, where James tried unsuccessfully to find employment in that town's shoe industry, both he and Josephine returned to Starrucca where Fred's father William was born. Shortly thereafter the Skinner family moved to Susquehanna, where James found occasional employment at odd jobs, mostly house painting. Fred's grandfather was a portly figure who sported a striking handlebar mustache. A man of few words and little ambition, Fred recalled of him that, "if he lived any life at all it was my father's."[13] He delighted in his son's eventual legal career and would attend William's local court cases, even after deafness left him unable to follow the proceedings.

Josephine was a small woman with a "catlike" face who limped slightly from a childhood accident. Clearly, her grandson did not find her attractive. He described her in a memoir as a woman who had whiskers, used too much makeup, had frizzy hair, and later wore a wig, which James referred to as "the transformation."[14] Her personality was no compensation. Though less phlegmatic than her husband, Josephine was pretentious and strained to conquer her lowly origins, usually unsuccessfully. Nervous and loquacious, she was remembered by Fred as "putting on airs" in front of him and his father.[15] She also liked to tell jokes, usually the same ones, many of which were scatological. Her grandson recalled her pleasure in repeatedly telling how she cleaned his toilet when he was a child. On one occasion, she "thought my brother and I should be 'wormed,' " a common purgative among the rural poor. Nor was she particularly adept in the home; her plants tended to die, and her homemade preserves exploded in their containers. "My grandmother was a fool," Fred said bluntly. He was "contemptuous of my Grandmother Skinner and scarcely less so of my Grandfather Skinner," and he believed his father's insecurity and unhappiness could be traced to Josephine's baneful influence.[16]

But others remembered Josephine as generous and grandmotherly. A neighbor wrote to Fred: "I adored her. She reminded me of my grandmother whom I only saw summers when I was sent to Kentucky. . . . Your grandmother said I came to visit her especially when her roses were in bloom. She always sent me home with an arm load."[17] One incident may have been crucial to Fred's negative opinion about his grandmother. When he was around ten years old, she opened her oven door and vividly portrayed to him the searing, everlasting hellfire: the wages of sin. She literally scared the hell out of him and he remembered having nightmares afterward.[18]

Fred was more inclined toward his maternal grandparents, Charles and Ida Burrhus, whom he and Ebbie called Mr. and Mrs. B. Charles grew up about fifty miles northeast of Susquehanna in Walton, New York. Like thousands of young American men who heeded President Lincoln's call in the early days of the Civil War, Charles enlisted in his state's volunteer infantry. He served several years in South Carolina and was with General Sherman on his famous march to the sea. During the mid-1870s he came to Susquehanna to help rebuild a washed-out bridge, and while there he met and married Ida Potter. Charles found work as a carpenter for the Erie Railroad and eventually became shop foreman. This mustached, rather squat man with brown hair that never turned gray spent nearly half a century in the Erie carpentry shop. Fred was amazed by his grandfather's ability to produce "marvelous points on my pencils . . . done with a pocket knife kept razor sharp—and he could peel an apple in one long unbroken string of reversing s's."[19]

Grandfather Burrhus had a certain flamboyance. Although he never owned a home, he purchased several new automobiles in an age when owning even one was exceptional. He urged his grandson to enjoy life and occasionally to break the rules, as he himself did when he put a few teaspoons of coffee in Fred's milk at holiday meals—a practice highly disapproved of by Fred's mother, who believed coffee would stunt her son's growth. But, to Fred, "coffee has never tasted that good since."[20] Neither grandfather, however, was especially attractive to young Skinner, and certainly not adult models to be revered.

The marriage of Charles Burrhus and Ida Potter joined two young people from respectable families. Ida's mother was one of the earliest settlers to the Susquehanna area, having arrived in 1829 from Vermont. Ida could trace her lineage back to a Captain Potter who had served under George Washington in the Revolutionary War. She was an attractive woman who wore steel-rimmed glasses and whose long skirts all but hid her black-buttoned shoes. Active in the Susquehanna Women's Auxiliary, Ida was an excellent cook and skilled at needlework. She was a great reader of fiction, a pastime none of Fred's other grandparents cultivated.[21] She was often unwell, and Fred recalled her emphasizing the shortness of life: "If I should say 'I wish it was Saturday' or 'I wish Christmas would come,' she would appear scandalized . . . and tell me *never* to wish away part of my life." An Erie Railroad metal worker by the name of Starkweather, whose daily contact with metal was believed to have given him a special power to heal, impressed Ida. She had small pieces of flannel on which this man had placed his hands, which she wore on her chest for bronchial trouble.[22]

After her death in 1923, a sealed letter was opened and read, as she had requested. With family members gathered, her son-in-law, William, read a few lines of it to himself and promptly threw it into the lighted fireplace. Eighteen-year-old Fred wondered whether it might have revealed a sexual indiscretion on the part of his grandfather. Later, he believed that both his father and Grandfather Burrhus had suffered from sexual frustration (a common early-twentieth-century complaint among the middle class), and that by burning the letter his father had been protecting his father-in-law "as a fellow sinner."[23]

BOTH SETS OF FRED'S GRANDPARENTS were frequent visitors at the Skinner house on Grand Street. Christmas Day was spent with Mr. and Mrs. B on Myrtle Street, and other holiday gatherings took place at Grandparents Skinners' home on Jackson Avenue. The three generations socialized as an extended family, but the social distance between the life-style of William and Grace Skinner and that of their parents was greater than one might have imagined in a small community like Susquehanna.

Born in 1875, William Arthur Skinner was positioned between two Americas, one rapidly fading into nostalgia and the other vibrant, aggressive, and still being shaped. His mother had been raised on a farm, but his son would spend most of his adult life in a metropolis. William would himself move to the larger community of Scranton, Pennsylvania, in his late forties. Unlike his wife and sons, he had been raised in a socially marginal family; neither James nor Josephine Skinner had the education, wealth, or family connections to make them socially attractive in Susquehanna. When William took his sons to family reunions, they visited a place with no indoor plumbing and with floors that went rugless and unpainted. This was a heritage to which no one in the Skinner family wished to return.

The only child of a lackluster father and a socially aspiring mother, William became the vehicle of his mother's dreams and ambitions. Josephine had pinched her son's nose as a baby "to make it sharper and more distinguished looking," but her grandson felt "she had pinched him in other ways, not so easily identified or described, and in the long run the other pinches were more painful and possibly not much more successful."[24]

After graduating as salutatorian of his Susquehanna High School class, William worked for a short period as a draftsman in the Erie Railroad Mechanical Engineering Department. He showed little mechanical aptitude, however, and decided in 1895 to enroll in law school in New York, supporting himself as a bookkeeper in his half-uncle's decorator store on Broadway. Having read some law while a draftsman, he was able to complete the two-year course in one year. William received a handwritten certificate rather than a diploma from The University of the State of New York. His son later commented that "it was not the kind of thing to frame and hang in an office."[25] The American middle class in the early twentieth century was self-consciously professional. William was in step with the times and on the road to professional standing—the era's conduit to social status.

William passed the bar examination in June 1896, the year of the great political struggle between representatives of agrarian and urban-industrial America. William McKinley's victory over William Jennings Bryan marked the passage of an older way of life and the gathering dominance of a new one. Will Skinner was on the side of America's future—so much so that he became politically active. During the rematch of Bryan and McKinley in 1900, the twenty-five-year-old attorney gave a highly praised speech for the McKinley-Roosevelt Club at Montrose, the Susquehanna County seat. Years later he would speak at political rallies throughout World War I. All his life, William Skinner remained a staunch Republican.

With his law degree, an office on Main Street, and recognition as "Lawyer Skinner," William quickly gained local prominence. His professional credentials rendered him attractive to Susquehanna's largest corporation, which, like those across the land, was becoming ever more dependent on the skills of professionally trained accountants and lawyers. In 1907 he was hired as an attorney for the Erie Railroad, the sure sign of a rising reputation. For social success, however, a promising young man needed to marry a suitable girl. Here, too, William Skinner succeeded. The *Susquehanna Transcript* proudly announced his marriage to Grace Madge Burrhus in April 1902: "the Bridegroom is a popular and rising young lawyer and the Bride one of Susquehanna's fairest and accomplished daughters."[26]

Grace, born on June 4, 1878, was the oldest of four children and three years William's junior. Only she and one brother, Harry, survived into adulthood. Grace's chestnut hair and shapely figure guaranteed her many suitors. She was also gifted with a beautiful contralto voice, which brought her local acclaim that encouraged the prospect of a musical career. Her first public appearance as a singer was for a Universalist Church benefit in 1896. She sang in a local group and also performed at Susquehanna's Hogan Opera and in engagements in various local communities. Grace saved all the newspaper accounts of her modest but memorable musical triumphs.[27]

Like her husband, Grace attended Susquehanna High School and also graduated as salutatorian of her class. To meet her professional goals outside music, she learned typing and shorthand and was hired as secretary to the mechanical superintendent of the Erie Railroad in 1901. But early-twentieth-century American women usually sacrificed their careers when they married, and Grace was no exception. Even though she still cared a great deal about her standing in the community, henceforth her status would be associated with her husband's professional position.

Grace was impressed by William Skinner's rising reputation as a lawyer and political speaker. He was not physically remarkable, nor had his family much to recommend it. Before their marriage William had once tried to hide his parents' shortcomings. At a family outing to which he had invited Grace, he also invited relatives from sophisticated New York City.[28] It is doubtful that he was able to convince her that his parents were not social embarrassments, but he did seem to offer her a promising professional future.

A young couple married in 1902 could look forward to raising their children in a prosperous, stable, and progressive time. The severe economic depression and widespread labor unrest of the 1890s had abated. The nation had emerged from the Spanish-American War as a world power. The new century opened with the prospect of astonishing technological progress. Electrification of utilities was making rapid strides, the automobile would soon begin its meteoric rise, and the Wright Brothers were on the verge of making the first manned, sustained engine-powered flight. Even American morality seemed to be markedly improving, as clergymen spoke of the coming of a Christian Brotherhood when violence and disputation would be but an unpleasant memory. Grace and William shared the ethos of this Progressive Era, which assumed that Americans could look forward to uninterrupted economic, political, and moral progress.[29] This exaggerated faith in a better tomorrow was a powerful source of social and personal optimism for the American middle class in the early years of the new century. William believed wholeheartedly in the progression of the generations, each succeeding one bettering the last, and this may have been a fundamental courting point with Grace. After all, her own father, though of higher social standing than James Skinner, did not have the commanding presence of a professionally trained lawyer.

Susquehanna's future in 1902 seemed the nation's writ small. There was a general mood of prosperity and well-being. The town boasted an amusement park, a skating rink, visiting circuses, a racetrack, a county fair, a ballroom, and an opera house. By 1909 Susquehanna had its first automobile, which later became the town taxi, and the following year William Skinner purchased a Ford. Grace had good reason to believe that her husband would be successful in such a vibrant local atmosphere.

Although William ran unsuccessfully for mayor in 1903 and for district attorney in 1904, he was elected president of the Susquehanna Board of Trade and appointed United States commissioner for the Susquehanna district. A few years later he became director of the Susquehanna Telephone and Telegraph Company as well as director of the First National Bank. He was also borough attorney and the leading advocate for sewage and pavement improvements on Main and Exchange streets in Susquehanna—the first modernized streets in town.[30] William Skinner was a "wide awake young lawyer" embarking upon a publicly visible career at one of the most optimistic moments in the nation's history.[31]

When Fred was born, on March 20, 1904, the *Transcript* proclaimed that "Susquehanna has a new law firm, 'Wm. A. Skinner & Son.' "[32] William loved to tell friends of the newspaper's prediction. With the joy of their first child and Will's bright career prospects, the Skinner family seemed poised for a wonderful future.

Fred recalled his father as a gentle parent who never physically punished him, preferring to express disappointment or to attempt good-natured ridicule instead. For example, he used to "slump across the room to show me how round-shouldered I looked."[33] But William's mimicking was not effective enough to work as a behavior modifier. Nor did he cultivate a strong bond with his firstborn. To some degree this resulted from their different talents. William lacked his son's mechanical dexterity, and even though he was a first-rate speaker, he did not have Fred's verbal facility or, as Fred would gradually learn, intellectual interests. The *Transcript*'s prediction could not take into account an increasingly evident quality about young Fred: his ingenuity and independent thinking.

There were, however, perks to be had by being the son of William Skinner. As attorney for the Erie Railroad, William received free passes for rail travel, and every Saturday afternoon the family would go to Binghamton. Fred fondly recalled these outings, "shopping for clothes at Weeds, seeing sepia-toned Thom. H. Ince movies, having supper . . . in a white-tiled restaurant and waiting in the depot for the #26 at nine o'clock" to return them to Susquehanna.[34]

There was another trait of William's that his son criticized, perhaps because to some degree he shared it: vanity. His father boasted of his accomplishments to peers and underlings—neighbors said it was always the "big I and little u" with Will Skinner—but never felt comfortable with social superiors, not knowing how to initiate appropriate conversation. "As he rose in the world," Fred recounted, "he found himself ill-prepared for each new step. Any assurance that he was successful was terribly reinforcing. . . . He listened for it, glowed under it. He often praised himself, obliquely or openly, and my mother was always there to protest."[35] William seemed to be a man under the control of both a socially ambitious mother, who had been unable to teach him the skills and confidence to achieve her aims, and a wife equally unable or unwilling to appease his powerful need for approbation. But he could be a sensitive communicator, as a friend recalled:

> I had one contact with your father. When I was around nine or ten, I was sent to pay the telephone bill to Lawyer Skinner. I must have told his secretary that I had to see *him,* and I must have refused to say why. . . . She finally opened a door, and there sat a lot of men around a table, and your father at the top. I finally divulged that I wanted to pay the telephone bill. Most men would have wanted to kill her or me, or both of us. Your Dad never turned a hair. He said something to her in a low tone, and I finally yielded up the money to her.[36]

Personal conceit notwithstanding, William Skinner did succeed as a local lawyer, political orator, and enthusiastic town booster. Yet Fred portrayed him as pathetic, an object of sympathy, or simply ridiculous. When William took up pyrography, an artistic hobby in which designs are burned onto materials, his son was not impressed with the results. William simply burned in the markings already designed, Fred said, and even then the effect was shabby.[37] He was contemptuous of his father's acceptance of "the philosophy of American business unanalyzed."[38] He also implied that his father lacked the political sense and intellectual talent to rise very far in his chosen profession. William often wrongly predicted Republican victories in presidential elections, but remained loyal to the party even after the Great Depression. His aspiration to be elected to a judgeship would go unrealized. And Fred considered his father's attempts at writing stories and poetry doggerel. But they suggested yet another trait Fred would share with his father: a susceptibility to sentimental love.

Grace's influence on and control of William, Fred believed, contributed substantially to William's ineptness and eventual unhappiness. "She had *consented* to marry my father, and there was an element of consent in her behavior with respect to him throughout his life," he wrote. Her condescension toward William was apparently coupled with a lack of sexual intimacy.[39] Fred determined that his father was "intrigued by pretty girls" but never got anywhere with it.[40]

Fred recalled his mother's frequent reprimands to William, given in an I-told-you-so tone that he almost always let go unchallenged.[41] "My Mother," Fred said, "was the person who set the style of the house and the standards."[42] One of those standards was neatness. Grace, frustrated in her attempts to get her son to hang up his pajamas, would scold him day after day. His solution to the problem at ten years of age is a telling forecast of his later ingenuity:

> I solved it by building a little gadget. It was a hook on a string in that little closet where I would hang the pajamas and the string passed over a nail and . . . came down in the doorway and there's a sign saying "hang up your pajamas." Now if the pajamas were on the hook the sign went up out of the way, but when I took them off at night the sign came down on the door and in the morning I got up, got dressed, I started to go out, there would be this sign there. I'd go back, get the pajamas, hang them up and the sign would get out of the way.[43]

Grace also policed sexual behavior. Overhearing his mother and her friends as they noticed two children in a neighboring yard exploring each other's bodies, Fred recalled her severe reaction: "If I caught my children doing that, I would skin them alive!" Fred "well understood . . . the unfortunate effect of early punishment of sexual behavior or early conditioning of negative responses in preventing normal sexual behavior in the adult."[44] He admitted his fear of being discovered masturbating.

Grace continued to take pride in her own attractiveness—although she would have been the last to call it that—right into old age. She was conscious of diet and constantly enjoined Fred to eat slowly, to chew his food well, and to follow her example of standing for twenty minutes after each meal—a practice she claimed enabled her to keep her figure.[45]

Like thousands of other early-twentieth-century American children, Fred and Ebbie (born in 1906) grew up in an atmosphere in which prohibitions and habits were elevated to a code of behavior that, when violated, was said to result in dire consequences to moral and physical health. This was especially true of the upward-aspiring middle class, who needed self-control to achieve and maintain social respectability.[46] But in the Skinner household that code was not always equally applied. Fred recalled his brother once getting away with an act for which he believed he himself would have been punished. The family was gathered in the library before an evening fire when Ebbie rose to go to the bathroom. He mistakenly went into the kitchen and pointed his penis down into the coal scuttle. When he came out of "this little fugue," he laughed hysterically at his mistake.[47] His parents ignored it.

Grace clearly passed on a powerful social code to her eldest son. "My mother was quick to take alarm if I showed any deviation from what was 'right,' " he explained. "Her technique was to say 'Tut-Tut' and ask 'what will people think?' "[48] Social policing instead of approval may have made Fred unusually sensitive to praise when it did come. Later his psychology would emphasize the crucial effect of positive reinforcement on behavior.

Grace Skinner also believed that helping others was a duty. She was president or chair of numerous local organizations, yet Fred did not believe his mother really enjoyed serving: "It was rather artificial, perfunctory. . . . I never sensed any joy in her serving." Fred believed instead that her reward was the group identity she felt.[49]

Community service was a common option for American women who had not pursued a work life or had given up their jobs, as Grace had done, for husband and family. A powerful motive for such altruism was also the community disapproval a woman would face for *not* helping. Grace's appeal to others' opinions may well have contributed to Fred's own fear of making social mistakes, his easy embarrassments, and his self-denigration. Then, too, there was a crucial distinction between doing something because you enjoyed it and doing something because it was your duty. Skinnerian psychology would emphasize "natural" as opposed to "contrived" reinforcers.

Generally Fred tried to disguise any ill will he felt toward his mother, but on one occasion during his teenage years it surfaced:

> She was overworked, possibly disappointed in my father, at any rate easily upset. She and I were in the kitchen one day quarrelling. She made some critical remark. I said something like "there are other people to whom that might apply." The "other people" was, I think, the unkindest part—I meant her, of course, and it was obvious. She turned toward me, opened her speechless mouth, raised both hands like claws in the air and came at me. I held my ground and she stopped before she reached me. She tore off her apron and dashed out of the room and upstairs.[50]

By implying that his mother did not live up to her own moral code, he exhibited his cleverness in exposing the hypocrisy and artificiality of her domestic policing. Maintaining her roles as wife, mother, and arbiter of morals, as well as her civic responsibilities, must have been a taxing burden. In this instance, her son discovered not only the double standard but the strain.

Yet to interpret Grace Skinner's effect on her eldest son as simply that of a domestic controller, driven by an obsessive fear of "what people will think," would be mistaken. She also projected a powerful romanticism associated with her music. His father had played the cornet as a young man, but it was Fred's mother's music that had the poignant effect on him. His memory of her singing and piano playing remained extraordinarily vivid. Once while rummaging he came across a copy of "Little Boy Blue": "My mother owned the poem set to music. I have not seen the music for at least forty years but I can get gooseflesh and a chill just by thinking the tune and a few phrases—*O the years are many, the years are long.*"[51] Fred's sentimental reaction to his mother's music may have indicated a strong need for more affection and less judgment from his parents, especially his mother. In later years, his love for Richard Wagner's compositions would often evoke an equally strong yet unexplained sentimentality.

Fred was smitten by his mother's romantic presence. Later he would date girls who shared her physical traits, even one who shared her name. While her music and her example of community service were positive influences, her prudishness, condescension, criticism of his father and himself, and deference to that abstract but powerful social controller, "what people will think," also left their marks. The Skinnerian concern with controlling an organism's behavior in "nonaversive" ways may have originated in his relationship with an aversive, controlling mother. "We may not be free agents," he said, "but we can do something about our lives, if we would only rearrange the controls that influence our behavior."[52]

Fred also learned about unfulfilled ambitions from his parents, as the careers of judge and musician lay beyond their reach. But something other than their personal failure may have been behind Fred's harsh judgment of them. As a childhood friend observed, "I think you and I had a fault in common. We were inclined to be ashamed of our parents because they didn't have our knowledge (!) and tastes."[53]

PART OF THE WIDER CULTURE that embraced Susquehanna was a Protestant culture. From it Fred acquired the desire to be kind to others and to behave well. The latter meant essentially being liked. Although Grace and William were members of the Presbyterian Church, they were not enthusiastically religious, refusing to go to the evangelists' revival meetings that most of their fellow Protestants eagerly attended. Despite Grandmother Skinner's vivid presentation of hell, by the time he was around thirteen or fourteen, Fred entertained serious doubts about the afterlife. At the time of the revival meetings, "an electrician who had not attended [them] was accidentally electrocuted. He was the father of a friend of mine and I strongly resented it when the evangelist referred to his death as punishment for not attending the meetings. . . . I saw suddenly the frail humanness in religion and I must have revolted, not quickly but over a period of years."[54] The cultural legacy Fred would carry forward from this "fundamentalist mercantile culture" was belief not in heaven or hell but in the Protestant work ethic.[55]

Another important element in Fred's early years was his brother.

Two-and-a-half-years Fred's junior, Ebbie was an affable child who enjoyed raising pigeons and playing the clarinet. Later he would be a valued member of the Susquehanna High School basketball team. Ebbie was more outgoing than Fred, and Grace and William treated him more leniently—and not only with regard to sexuality. Once Ebbie found William's revolver and accidentally shot a hole through a bureau. His parents never criticized him for playing with the gun; they were just overjoyed that he had not hurt himself.[56] A fun-loving boy, Ebbie would tease Fred by repeating everything he said, which left Fred feeling powerless.[57]

Ebbie had the social grace and ease that Fred lacked. He fit the expectations of the conventional middle-class way of life, the one his parents most appreciated. Bright but not intellectual, he preferred school athletics to school itself, having fun with the guys to making things or reading. In all likelihood Ebbie was an easier child to raise than his older brother, more easily controlled, less self-centered, and less inventive. He was clearly the favorite.

But Fred was not jealous of his brother; indeed, he liked him, and it would be wrong to see them as taking separate paths right from the start. They played together; they enjoyed each other's company. They discovered sexuality, or rather Fred had discovered his and wanted to impress his brother. He remembered proudly displaying to him an erection, discreetly camouflaged behind a shower curtain.[58] But in time their interests diverged. Ebbie shielded Fred from the overweaning attention he would have continued to receive had he remained an only child. Indeed, after Ebbie captured his parents' affection, Fred began more and more to devise his own amusements, his own adventures and way of life. And when Ebbie entered high school, proved his athletic prowess, and extended his easy popularity, Fred felt even less a part of his parents' world. Obviously his brother was doing the kinds of things that most people in town, including Grace and William, expected a young man to do. Little wonder he earnestly sought an environment as well as companions to encourage what he enjoyed. What Fred Skinner enjoyed was making things, succeeding academically, and enjoying perks from a special teacher who shared with him her intellectual interests. The strictures of Susquehanna's environment did not contain Fred, although they marked him forever. Instead, they led him toward the freedom and adventure of other environments that would offer other experiences.

FRED INSISTED THAT he "was not born with a character trait called curiosity or with an inquisitive spirit or an inquiring mind." Rather, he gravitated toward a world that "richly reinforced looking, searching, investigating, [and] uncovering."[59] The influence of such a childhood would be lasting. As a childhood friend observed, "you were very sensitive, naive and . . . inclined to experiment."[60]

Fred's childhood inventions and activities were enough to turn a mother's head gray with worry and distress, not to mention the mess and the complaints from neighbors. As he was to tell it later:

> I was always building things. I built roller-skate scooters, steerable wagons, sleds, and rafts to be poled about on shallow ponds. I made seesaws, merry-go-rounds and slides. I made sling shots, bows and arrows, blow guns and water pistols from lengths of bamboo, and from a discarded water boiler a steam cannon with which I could shoot plugs of potato and carrot over the houses of our neighbors. I made tops, model airplanes driven by twisted rubber bands, box kites, and tin propellers which could be sent high into the air with a spool-and-string spinner. I tried again and again to make a glider in which I might fly.[61]

Sometimes his inventions went awry. A friend recalled an incident at her grandmother's house: "It seems some of your ammunition (a carrot, I believe) catapulted thru an upstairs attic window instead of over the roof tops to Erie Avenue as intended. . . . You came the next day . . . and you replaced the window."[62]

With a friend he strung wire on backyard fences and made a workable telegraph. He built a miniature theater and remembered "the satisfaction of arranging strings in such a way that the curtains attached to them parted with a single pull and closed with another." He fashioned toys and ornaments from papier-mâché. At a summer Chautauqua—a week-long extravaganza featuring traveling lecturers, musicians, and magicians—he was fascinated by a magician's ability to command balls on a track made of parallel rods to go up to the end of the track and then return: "The next day I made a similar device—and it worked. The balls were simply of different weights and responded to slightly different centrifugal forces or slopes of the track."[63]

One summer a magician/scientist came to town in the annual Chautauqua. Skinner recalled three "experiments": one involved the fusion of two large nails; in another, a stack of wood boxes was made to collapse when a tuning fork was struck nearby; and the last revealed a powerful gyroscope riding on two wheels on a horizontal tightrope. Nothing in Fred's fascination with this magic led him to the conclusion that he was predisposed toward science.[64] He was simply enthralled by these staged events and loved the gasps of disbelief and appreciation in the audience. He was just as delighted with the dramatic puns, parodies, and virtuosity of the Chautauqua musicians when *The Mikado* and *The Bohemian Girl* came to town.

One Halloween he constructed a device that made a loud buzzing sound and deposited it on the windows of neighborhood houses, running "like hell" while the inhabitants wondered what in the world was causing the noise.[65] For Fred Halloween was always tricks rather than treats.

He also made various musical instruments: something called a Willow-Whistle; kazoos from combs and toilet paper; cigar-box violins; noise-makers from spools and string; and various devices made of strings and buttons that could be twisted and released to produce a vibrating sound.[66] With a high school friend he invented a game, similar to Ping-Pong, that they called Teno-Ball. The balls they used did not quite have the bounce to keep the game interesting, but the boys were optimistic, nonetheless, going so far as to print a four-page pamphlet of rules and to copyright the name.[67]

Fred's inventiveness ranged beyond building things, beyond even the physical world itself. He once tried to determine whether the much-admired religious adage "faith will move mountains" actually worked, by practicing levitation and standing on a beam from which scales were suspended (a Fairback beam scale) and trying to make it tilt. He also had daydreams of being able to fly "usually in order to astonish people." Because many people believe that staring at someone's head long and hard enough will cause the person to turn around, "I tried it—with, of course, an occasional 'success' to keep me trying."[68]

The countryside surrounding Susquehanna, with its uncultivated, quasi-primeval ambience, encouraged roaming and foraging. At fifteen Fred and four friends went on a three-hundred-mile canoe excursion to Harrisburg. Fred dammed a creek to make a swimming hole where the boys could swim—along with a poisonous snake. He trapped and ate eels from the Susquehanna River.[69] He tramped the countryside and learned to identify local flowers, fruits, nuts, and berries, returning home with large quantities of arbutus, dogwood, honeysuckle, and skunk cabbage, as well as more edible delectables: apples, peas, cherries, hickory nuts, chestnuts, gooseberries, raspberries, and currants. These bountiful harvests were not necessarily dumped and forgotten. A neighbor recalled Fred "direct[ing] Grand Street kids in making apple jelly from sour green apples." After cooking the apples for a considerable time, Fred declared, " 'This won't gel but we *can't waste* it—Let's drink it!' "[70]

Local animals and their behaviors also interested him. He caught bees in hollyhock blossoms, watched cows being milked, and looked on while bulls or dogs copulated. With his best friend, Raphael Miller, the son of the local doctor, he once tried to make pigeons drunk by giving them alcohol-soaked corn. (Could they have been Ebbie's poor pigeons?) He observed in amazement while a relative killed chickens for Sunday dinner and the animals ran a few steps after being beheaded.[71] Inspired by reading about how to make money in furs, he purchased and set some traps, but he never caught anything. On numerous occasions he returned home with turtles, chipmunks, or other local animals that his mother probably did not enjoy having about the house.[72]

Whatever the controls his parents exercised, Fred had considerable physical freedom to explore, to observe, to tinker, and to invent. Like Huckleberry Finn, who was cautioned by Widow Douglas and Miss Watson but who lived an adventuresome life nonetheless, Fred countered control in one area with a free inventiveness in another. The juxtaposition of strong domestic controls and generous physical freedom seemed to suit him exceptionally well.

He enjoyed intellectual as well as mechanical pursuits. On the lighter side, he read Buster Brown, the Katzenjammer Kids, and Tom Swift. His father, an easy mark for book salesmen, purchased volumes with ambitious titles such as *The World's Great Literature, Masterpieces of World History,* and *Gems of Humor.* William also had a series of books on applied psychology, which impressed young Fred only by their lovely bindings and one esoteric example of bad psychology—"an advertisement for chocolates showing a man shoveling cocoa beans into a large roasting oven"—instead of eating chocolate.[73] As a boy the only kind of psychology he read was self-help, which in those days taught what behaviors were required for conventional success and morality. He did not derive his keen interest in the control of organisms from these cultural commands and restraints, but when they were implemented by his parents they made a powerful impression. He escaped not only by inventing things but by constructing an alternative intellectual world.

Fred also liked Little Books, which were then popular with young readers. He had a tiny dictionary (only slightly larger than a postage stamp) and dozens of other very small-sized books. Owning Little Books gave him a special power: "One could contain vast treasures in two cupped hands—a sort of literary miserliness was encouraged, a sense of personal possession. Knowledge may not have been secret, but it was easily secreted."[74] By keeping and reading these books, he began to learn there was an intellectual realm not only different from but distinctly superior to the one his parents inhabited:

> The books I had as a child belonged to two worlds. One was the world I lived in—the books my parents read to me, about people and animals and things others talked about. The other was a foreign world. The illustrations were different and better. The texts were more grown up. The print was different—even the bindings. I suppose there were only a few books in that world but it still seems to me clearly defined.[75]

Coveting publications such as the Little Books gave Fred the excitement of discovering things independently of his parents: "One book in the other world was about a war of the animals. It developed the theme that the lion was the king of beasts. I was convinced. No divine right could have better established the lion's legitimacy."[76]

Moreover, he built a private place where he could read. When he was about ten, he introduced himself to the world of boxes: "Certainly 'a box to hide in' is something most of us have wanted at one time or another. . . . For some reason or other this seemed to be the right place to go when I felt like writing something."[77] His initial "box to hide in" was fashioned from a packing case into which he would crawl. On a tiny shelf he kept a pad of paper and a pencil. Other boys in the neighborhood built small shacks or hid in holes they had dug, but Fred's box building seems to have been a more sophisticated and private enterprise. He added a curtain that could close off the opening and small shelves to hold his books, writing materials, a candle, and the like.[78] Here was a boy's study, a place separate from the Skinner household, where he could intellectually detach himself from parental guidance—in a special sense, the first Skinner box. One wonders what his parents thought or said as he sat for hours in his cubicle. It was another way of making do, inhabiting this small, private space designed to facilitate his concentration on these beautiful, fascinating little books—a home within the home from which he was becoming ever more disengaged. As an adult, Skinner always highly valued his place for thinking and writing. And he did most of his intellectual work at home in a private study rather than at an academic office.

MUSIC WAS ALSO IMPORTANT in Fred's boyhood, and not only his mother's sentimental music or the family Victrola. He took piano lessons from "Harmy" Warner, the Presbyterian Church organist, "an old man who sucked Sens-sens" and taught him to "spell *cabbage* on a [musical] staff."[79] In high school he played the saxophone with a group called the Susquehanna Erie Railroad Band, which was to tour neighboring towns during World War I selling Liberty bonds.[80] Fred never learned to play more than a few bars on either the piano or the saxophone without sheet music. Nonetheless, he "made do" with music as he did with his inventions and experiments and took as much enjoyment in it as he did in making a backyard telegraph or a carrot-shooting cannon. His friend Ward Palmer, whose father was an auto mechanic and whose mother was believed to have family connections with Frank Lloyd Wright, provided further musical enjoyment. After playing tennis with Ward, Fred would go to Ward's house where they spent considerable time listening to Palmer's extensive collection of opera records, including Wagner, who was to become Skinner's favorite composer.[81] A friend recalled Palmer's generosity in allowing interested neighbors to spend many contented hours listening to his music. Palmer also had a miniature stage on which marionettes performed operas.[82]

Perhaps because of Fred's and his mother's love of music, William Skinner took advantage of his free rail passes and treated the family to a performance of *Carmen* in New York City when Fred was in his early teens. The occasion, however, was marred by the embarrassment that Fred and Ebbie felt in their new tweed suits, which marked them as socially backward.[83] The fact that his parents did not understand what a young man should wear to a New York opera did nothing to enhance Fred's opinion of their social grace.

With his mechanical aptitude and love of nature, music, and reading, one might guess that Fred found school boring. But he liked it and described the Susquehanna public school as "small, serious and good."[84] Unlike many American inventors, Skinner had considerable verbal as well as mechanical ability. Generally, this was an academic asset, but on occasion his verbal dexterity got him into trouble. Once in an eighth-grade science class, "We were discussing fatty acids. One of the more buxom girls in the class was at the blackboard. I whispered in a loud tone to another boy *'There's* a fatty acid!' Miss Keefe took the matter up with the principal."[85]

Fred's twelve school years were spent in the same small brick building on Laurel Street. He was one of eight graduating seniors, and he uncannily repeated the performance of both of his parents by graduating salutatorian of his class. He recalled gaining strong mathematics training there and learning enough Latin to read "a bit" of Virgil. Even though the school left him less well off with regard to science, he did not feel shortchanged because of the many physical and chemical experiments he did at home.[86] He was a good student in all subjects.[87]

Far and away his most important intellectual influence at school came from a special teacher, Mary Graves. The daughter of a local stonecutter-turned-amateur botanist whose belief in evolution branded him the town agnostic, she maintained a high level of cultural interest that impressed Fred.[88] She did her best to keep Susquehanna's public library up-to-date; she was one of the founders of the local women's literary society, the Monday Club, to which Grace Skinner belonged, and she taught Sunday school at the Presbyterian Church near a stained-glass window that had been contributed by a family named Frazier. Years later Skinner returned to the church and guessed that Frazier, the protagonist of *Walden Two,* was originally inspired by the many hours he spent sitting near that window.[89] Although not an agnostic like her father, Mary accepted Darwinian theory, and her treatment of the Old Testament was metaphoric rather than literal. She was Fred's art teacher in grammar school and English teacher in high school, and her enthusiasm for both these subjects infected Fred. In high school, he was invited to Miss Graves's home, where they discussed literature and science. She lent him books. Her death from tuberculosis shortly after his graduation touched him deeply.

More than any other individual in Susquehanna, Mary Graves helped Fred see the limits of his parents' intellectual world. Her influence was the strongest during his adolescence, a time when dissatisfaction with one's parents is often at its zenith. In many ways, Mary became the reinforcer while Grace remained the enforcer; the latter encouraged compliance, the former independence. Mary recognized and praised Fred's intellectual abilities, abilities his mother largely ignored and on occasion found troublesome. Rather than being a domestic policer, Miss Graves was an educational catalyst who encouraged Fred's natural curiosity and urged him to find answers for himself from books and nature. Intellectual independence became essential to Skinner's behavior as a scientist, and Mary Graves nurtured that practice in him. She kept a small notebook in which she carefully described plants (she, like her father, was a botanist) as well as Darwinian and religious observations.[90] Skinner's later practice of keeping a notebook, even his reliance on a cumulative record of behavior, may have in some exemplary way benefited from Miss Graves's attempts to record her thoughts and observations accurately.

A memorable example of Miss Graves's championing of intellectual independence, although perhaps to her own embarrassment, occurred in Fred's eighth-grade literature class. Fred mentioned to his father one evening that his class was reading Shakespeare's *As You Like It.* William told Fred that Francis Bacon was the real author of Shakespeare's plays. Deferring to his father's authority, or perhaps playing devil's advocate, Fred triumphantly announced to his class the next day that Shakespeare was a phony. Miss Graves challenged William Skinner's intellectual credentials and advised Fred to find out for himself who the true author was. Searching the Susquehanna Library, he discovered Edwin Durning-Lawrence's *Bacon Is Shakespeare* and told the class of his finding the next day.[91]

His library research had an unexpected dividend. Fred became interested in Francis Bacon and began reading biographies, even attempting to read Bacon himself, delving into *Advancement of Learning* as well as the classic treatise *Novum Organum.* He did not remember becoming a Baconian at the time, but later adopted Bacon's dictum that to be commanded, nature must be obeyed. Skinner emphasized that he became a Baconian also with respect to scientific method, education, and the "abiding principle that knowledge is power."[92]

As Fred neared graduation from high school, Susquehanna, despite the salutary influence of Mary Graves, provided him with fewer and fewer opportunities to express his expanding intellectual interests and independence. He became more uneasy with the oppressive parental and community controls that Sinclair Lewis had immortalized as the tyranny of the village in *Main Street* (1920). Like Lewis and other American intellectuals, Skinner would eventually rebel against the village, but the village would never entirely leave him.[93] The ethic he inherited from this "shabby fundamentalistic, mercantile town" dogged him all his life. Small-town life did not necessarily smother all intellectual curiosity nor prevent imaginative improvisation, but it presented obstacles that blocked or stunted the kind of future he was seeking: "People know each other and have done so over long periods of time. . . . They are all of a sort of common police force, censuring, commending, keeping in line. . . . This general policing has its price. . . . Conformity is costly."[94] Young Fred Skinner did not plan to become a scientist—let alone a behaviorist—so much as he became ever more aware of the kind of life he did not want; and he did not want to stay in Susquehanna with his parents.

William and Grace could approve or disapprove, but they could not stimulate, guide, or wholly empathize. As he would describe them: "They could not say that a person, friend, colleague, or short story was good. They could not evaluate an experiment. And in giving them up as sources of praise, I never found or even sought a replacement. Hence my failure to make contact with the psychology of the time. Hence, thank God, my chances of making contact with the psychology of the future."[95] The psychology of the time was the psychology of the Protestant work ethic, the psychology of sexual restraint, the psychology of doing one's duty for fear of what others would think. It was for young Fred Skinner a code of behavior rather than a psychology in any scientific sense. But his failure to be reinforced by his parents, or their way of life, simply made it possible for him to become something they were not. And as his enthusiasm for Ward Palmer's music and Mary Graves's intellectual world revealed, he was beginning to be attracted to an alternative culture. One culture would have to replace the other before the psychology of the future could replace the psychology of the past.

Growing alienation from his parents, broadening intellectual interests, as well as a quick, probing intelligence gained Fred a reputation for opinionated arrogance. As one of his few intellectually inclined Susquehanna friends, Annette Kane, recalled in a letter to him: "We just had so many ideas and were so hell-bent on defending them, that when we met, an argument sparked, and usually heated up in no time. . . . I remember one day you said you would like to write like Dostoievsky, and, to torture you, I said I'd like to write like P. G. Wodehouse. You were so disgusted, you got right up and went home."[96]

Fred's high school principal, Professor Bowles, who was also the mathematics teacher, took a special interest in him. Bowles, a devout Catholic, once lent him a polemic against evolution called *God or Gorilla.* Just before his graduation from high school, Professor Bowles took Fred aside and told him: "You were born to be a leader of men. I just want to say one thing. Never forget the value of human life." Fred was stunned, since he knew no one saw him as a leader among his peers. His brother showed more signs of leadership.[97] Bowles's comments may simply have been the recognition of a bright student with a high energy level for whom he wished these qualities directed toward a conventional, constructive end. He was no doubt worried about Fred's radical intellectual leanings, saw his intelligence, and wanted him to come back into the fold.

But Skinner had little time to ponder what his principal may have meant. The last two years of high school were especially busy. Apart from school, he was working for a shoe salesman on Main Street, reporting and writing for the *Susquehanna Transcript,* plus playing the saxophone in his band two nights a week.[98] The violinist of the band recalled how they used to accompany silent movies, by the end of which Fred's teeth would be loose from playing so much.[99] Even having fun was hard work.

Fred also worked hard at having fun with girls, but remembered himself as sexually inept. One romantic interest was an Irish Catholic girl who, it seemed, did not return the interest. For a time he dated the daughter of the local barber. She worked in the ice cream parlor after school and Fred became a steady visitor there. He was allowed to touch her above the knee, but no higher; she had a strict code that permitted suitors, depending upon their rank, certain sexual privileges. Fred did not rank very high. But the big love of his high school years was Margaret Persons, whom he dated during his senior year: "Every Sunday after dinner I would comb my hair, walk down Church hill to the Sugar Bowl on Main Street, . . . buy half a pound of milk chocolates, walk out to the far side of West Hill and knock on Margaret's door." On these Sunday evenings they would either take walks or Margaret would play her mandolin to Fred's piano accompaniment in the parlor. Fred recalled trying to improve her playing: "She was the first who suffered from that."[100] The relationship ended when the Skinners moved away from Susquehanna.

After the Great War, Susquehanna went into a depression. Changing railroad technology, especially the replacement of the steam engine with the diesel motor, forced the Erie shops to close. Population declined. Business failures and suicides grew in number, one being Fred's old shoe-store boss, who hanged himself when his business failed. Banks refused loans. The local library closed. William Skinner was deeply troubled by these events and tried to boost the town's fortunes by urging the governor of Pennsylvania to consider the gassification of coal—a technological innovation that would particularly help the Susquehanna area. But his plan went unheeded, and his own position as attorney for the Erie Railroad stagnated.

William had badly damaged his local reputation years earlier by unpopularly defending an Italian strikebreaker during the general machinists' strike in 1907. The Italian was accused of murdering a striking Irish worker. William's defense of him in a predominantly Irish, English, and German town probably ruined any prospect for political office, especially the judgeship he so wanted. And as the Erie Railroad and town declined, it was increasingly difficult to find profitable cases if one was not associated with a large law firm. William and Grace's faith in progress seemed, metaphorically speaking, to have derailed.[101] Indeed, the whole nation's progressive mood deflated. Once again, Susquehanna seemed America writ small. A postwar economic depression, widespread strikes, unreasoned fear of foreign radicalism, political retrenchment, reaction, and cynicism replaced the ebullient progressive ethos of the previous twenty years. Having come to expect a continued journey of social and personal improvement as natural, the Skinners, like thousands of other Americans, were bewildered when it slowed.

An unexpected opportunity arose in 1922 for William to become junior associate for the general counsel of the Hudson Coal Company in Scranton. The salary was considerably higher than it had been even in the best Susquehanna years, and he had no compunction about being a company man, especially when prospects for promotion to senior counsel seemed excellent.[102] Grace looked forward to a new house, a maid, a more sophisticated social scene. The popular Ebbie would quickly adjust to a new school and a new town. Fred was also elated. He was just finishing high school and was ready for a change. He remembered reading a short story by Francis Noyes Hart titled "Contact!"—a word used by World War I pilots before takeoff, which depicted his mood. "College and Scranton meant a new World!" he wrote in his autobiography.[103]

It had always been assumed that he would go to college. The son of a professional man could do nothing else; besides, Fred wanted higher education. He aspired to a career in creative writing. Neither he nor his parents had any particular school in mind when a family friend recommended Hamilton College in Clinton, New York. Hamilton's admittance standards were undemanding, requiring only a certificate from an approved high school and "a satisfactory testimonial of conduct and character."[104] Principal Bowles sent a recommendation to Hamilton, noting Fred's special aptitude in mathematics, English, and history and describing his study habits as diligent and thorough.[105] The testimonial was provided by a Susquehanna alumnus of Hamilton College and sent to Hamilton's president, Frederick C. Ferry. Young Skinner, it related, was a "willing worker and a conscientious student" whom "you cannot afford to turn down." However, the writer added,

> It is only fair . . . to catalog some of his bad traits as well as the qualities in his favor. Frederic is passionately fond of arguing with his teachers. He is quite a reader and although I do not think he actually supposes himself wiser than his teachers, I have found him [to give] that impression in extemporaneous debate. These debates are frequent for he requires a reason for everything and mere statements with no proof never find a ready believer in him. When he is engaged in a heated debate, Frederic is apt to resort to sharp or bitter retorts. This has lost several friends for him in the past, friends who failed to consider that the expression was stronger than the thought.[106]

Hamilton accepted him despite the warning. On the day he left Scranton for the three-hour train trip to Clinton, his parents were not home. He wandered the house and finally visited a nearby grocer, who was not the least bit interested in his going away to college. "In Susquehanna everyone would have been interested," he mused. Nonetheless, he was in a state of "uneasy joy." "I did not know what lay ahead, *but I was getting away from my parents.*"[107]

SKINNER CARRIED TWO CONTRADICTORY American legacies with him on that train to college. One was freedom and the other control. One was making do, invention, improvisation, and intellectual investigation; the other was conventional, polite acquiescence to social codes. One was a heritage of disorder and messiness, the other an ordered progress. Writing to a childhood friend years later, he said: "I loved my life in Susquehanna in spite of the grime and disorder. . . . It had a good culture."[108] He believed that small-town, face-to-face culture had been important in his life, shaping his capacity to explore. It had also taught him about the class system, "groups of people with their own cultures, marked off in some strange way from each other"—Catholics and Protestants, Irish and Italians, mothers and fathers, parents and children, conventional citizens and eccentric intellectuals.[109] Social and intellectual distinctions mattered a great deal in Susquehanna and would continue to be crucial in his future. The prevailing social codes of "civilized morality," the "progressive ethos," and the "Protestant ethic" also marked him and, as he later judged, not always negatively: "Although I was not particularly happy about my childhood background I now see that it drilled enough of the Protestant Ethic in me to permit me to put up with many aversive features of my educational background and to get from it what I really needed to be, an independent scholar."[110] From Susquehanna, situated in a progressive and relatively innocent early-twentieth-century America, he learned to want a world "to be so good everything you do is reinforced, the things you make are nice and you're glad you made them, the friends you have are nice and you're glad you have [them]."[111]

But the time had come to "make do" in another environment, another culture. He was both relieved and exhilarated to escape his parents, though he could not yet articulate what he had escaped, and later admitted he had not in many ways escaped Susquehanna at all.

2

Between Two Lives

At Hamilton College I was between two lives. I was prepared to be one kind of person and turned out to be another.

—Interview with B. F. Skinner, August 13, 1990

When eighteen-year-old Fred Skinner arrived at Hamilton College, his environment shifted from a railroad town to a college village. Clinton, New York, was founded in 1787 as a dairy farming village but had long been associated with higher education. Yet a curious similarity existed between these two American towns: their proximity to mid-nineteenth-century perfectionist movements. A few miles from Susquehanna, Joseph Smith had written the Book of Mormon in 1819; and Clinton, New York, was less than twenty miles from Oneida, where John Humphrey Noyes had established a utopian community in 1848.[1] Skinner would later remark that each of these nineteenth-century perfectionist examples showed how "you could step in and do something about your life."[2]

Hamilton College was founded in 1793 as Hamilton-Oneida Academy and was chartered in 1812 as Hamilton College, a men's liberal arts school named after Alexander Hamilton, a member of the academy's first Board of Trustees.[3] Entering freshmen in 1922 paid an annual tuition rate of $150, reflecting a recent $30.00 increase. A week's board at the Hall of Commons was $6.50 and included breakfast rolls, which were "tossed from one table to another"—at times as if they were "lethal weapons."[4] Here, for four years, Fred would live in a more traditional America, a place where old wealth and family reputation counted more than the progress of the new railroad culture.

Unlike many Hamilton students, whose fathers or brothers were alumni, Fred was entering a world where no personal or family reputation preceded him. Yet the young man who disembarked from a taxi in front of the Chi Psi fraternity house that September was confident that he would excel academically and socially in his new setting. A letter to his parents was both reassuring and slightly condescending:

> We got to Clinton . . . and . . . to the Chi Psi (pronounced Kye Sigh) Fraternity house. . . . After having dinner at the Chi Psi's I went for a room in a dormitory and got one in North Hall. This is the oldest Hall here and of course not very modern. It is, they say, the best heated and has the best showers. . . . I had a wonderful meal at the Beta Kappa House tonight. . . . The meal was great and served with the neatness and care of a fine hotel.[5]

An older Hamilton student recalled that "Fred was rather reticent about talking of his family. . . . [and] he wasn't too sharp in the market. I sold him my tiny roll-top desk which had cramped me for a year and at my original purchase price."[6]

Fred was one of 111 entering freshman, a radical difference from the 8 seniors who constituted his graduating class from Susquehanna High School. President Frederick Carlos Ferry welcomed them and urged them to attain the goal desired of every Hamiltonian graduate: "the ideal of the well-rounded man who has been exposed to the benefits of liberal arts learning."[7] Ferry, who had assumed his office in 1917, was also trying to raise faculty salaries, decrease the faculty-student ratio, build new faculty housing, and expand the college's physical equipment. He was a progressive, no-nonsense president, one who, to some extent, identified more with modern American business culture than with the traditional Hamilton environment.[8]

To help its charges achieve the goal of a well-rounded man, Hamilton College placed great emphasis on writing and public speaking. Throughout their four years, students were required to spend three or four hours a week preparing and delivering oral presentations.[9] The ideal Hamilton graduate could easily become a lawyer or a clergyman. Fred's freshman courses included English composition, algebra, trigonometry, intermediate French, modern comedy, elementary Greek, general biology, freshman declamation, and elements of public speaking.[10] He "hated" biology[11] and found oratory presentations frightening, as the novice speaker tried "to remember not only the words but the gestures appropriate to the issue at hand. . . . [The] initial terror [was] enhanced by pennies hurtling down from sophomores in the balcony, if one hesitated between sentences."[12] This was not the only humiliation young Skinner would face.

He discovered he was not nearly as sophisticated with language—the college's forte—as he had thought himself to be. Susquehanna teachers, even Miss Graves and Professor Bowles, may have on occasion disagreed with him, but they had not criticized local verbal customs. At Hamilton, however, Fred's speech teacher immediately took exception to his ending sentences with the word *up,* as in, "I cleaned up"; "I wrapped a package up." Fred also pronounced words ending in *-dous* as *-jous: tremenjous, stupendjous.* And he said *forhorrid* for "forehead" and *crick* for "creek."[13] Being caught in such slips mortified a young man who was easily embarrassed and craved praise in his new environment.

One mistake occurred in Professor Paul Fancher's English composition class. A student had used the expression "very interested." Instantly "my hand shot up. 'I was taught that you must say *very much interested,'* " Fred proclaimed. Later he noted, "It was pretentious. . . . It was a plea in my own interest against the class." Although no one mocked him, "I knew then, and I smart for it still, that I had turned the class against me." He blamed his small-town background for the fact that he "knew nothing of the levelling practices which keep members of a larger group in line."[14] Like the behaviorist he would become, he was acutely aware of other people's behavior as well as his own.

Other slips were due to simple naïveté. On a Saturday trip to nearby Utica for a haircut, he left without tipping, to the barber's astonishment. He had never tipped barbers in Susquehanna and was ashamed to learn from his roommate that it was customary. The same roommate also mentioned that it was not necessary to grind a cigarette to pulp with one's heel to extinguish it. This matter-of-fact criticism of a fashionable habit deflated the glamour of enjoying his favorite smokes, Pall Malls.

Fred's attempts to appear sophisticated and the resulting shame when he failed might have had less force if the Hamilton upperclassmen had not seen freshmen as the perennial source of social ridicule. Freshmen were marked men. Called "slimers" and required to wear green beanies on campus at all times, they had to respect all betters and could not exit a college building before sophomores and upperclassmen. They were fair game for the vicious pranks of upperclassmen, who abducted them, often taking them into the country and abandoning them there. There was "a general meanness displayed . . . by upper classmen" that fostered feelings of inferiority and social ineptitude among freshmen. Hazing was common in the fraternities that dominated the college's social life. The bathtub gin often provided at their house parties increased the chances for mischief and harm. President Ferry worked to abolish vicious practices such as the "gym show," in which freshmen were forced to strip and slide on their stomachs across a floor awash with a mixture of water and cornmeal until "the floor became bloodied."[15]

But Hamilton tradition dictated that freshmen pledge a fraternity and here, too, Fred made a social mistake. He joined Beta Kappa, only a local fraternity. Although several years later it would become a chapter of the national, Lambda Chi Alpha, the affiliation was not prestigious; Beta Kappa did not pledge the best athletes or boys who had social standing.[16] The choice of a "wrong" fraternity revealed Fred's naïveté as well as showing that his parents were of little social help. Nor did he have a close friend or mentor to guide him. Years later it still bothered Skinner when a Hamilton graduate would ask him what crowd he had belonged to. Perhaps, he acknowledged, he had taken on his mother's social aspirations.[17] At Hamilton, "there were 'crowds' I didn't belong to. I would feel very much an outsider entering the D[elta] U[psilon] house even for a few minutes."[18]

He also experienced a new physical isolation. Hamilton College sat atop College Hill, or "the Hill," as it was called, just above Clinton. And although the town was approximately the same size as Susquehanna, it was less commercial and was dominated by college administrators, faculty, and alumni. Outside the chore and routine of schoolwork, there was little to do, made worse by the Hill's distance from town. A fellow freshman recalled that the isolation of the place led students to go "stir-crazy."[19]

Fred found it more difficult to "make do" here. There was no garage workshop or messy railroad town to encourage tinkering and inventing, the hands-on activities he loved. Moreover, he did not make close friends his first year. A classmate later described him as a blue-eyed young man with "sandy hair that was usually awry. He was thin but wiry and very quick in his movements, speech and obviously, his thoughts. . . . His outstanding asset was a hearty laugh." Though initially he impressed one fraternity brother as "outgoing, friendly and appreciative of suggestions and advice," he developed a reputation as an outsider and was considered aloof, intellectual, and conceited.[20]

Later in his freshman year, boarding at the Beta Kappa dormitory at the bottom of College Hill, he would trudge up and down the steep slope to classes and campus activities several times a day, climbs for which the "city of stairs" had well prepared him. Once in class he discovered his teachers were not nearly as helpful as those in Susquehanna, who, with their small classes, had been exceptionally devoted. In high school his intelligence had been appreciated and highlighted by the attention of the beloved Miss Graves. Fred was certain his intellect would serve him well in college, but at Hamilton he learned to his dismay that exhibiting intellectual superiority was often ignored and even ridiculed. He recalled that "a good many of the people I knew . . . really didn't give a damn about getting a college education. They just did it. They weren't intellectually excited about anything at all."[21] Classmates chided and teased him about what in high school he had considered great intellectual discoveries.

He also found less esteem toward teachers. Students deferred to professors in class but did not necessarily respect them. Each professor, some with affection but others with ridicule, was given a nickname: "Stink," "Smut," "Swampy," "Bugsy," "Brownie," and so on. Fred found himself among typical American male college students of the 1920s—fun-loving fellows nervous about passing their courses but more interested in the fraternity social calendar and the next football game or tennis match than in their books.

The college also had rigid regulations. Students were required to attend chapel daily. There were few excused absences, and professors penalized the boys for being late to class. A bell tolled twelve times at the top of each hour and everyone was to be seated by the twelfth ring. Across campus could be heard shouts of "Hold that bell" as students scurried to make the deadline. Physical education was compulsory, presenting the less athletic, like Fred, with real dangers, as they were bumped and battered unmercifully in soccer and hockey. These requirements coexisted with the honor system for examinations. Students signed an agreement not to cheat on tests and papers. In return, professors allowed breaks during exams, during which questions and problems could be discussed so long as answers or hints were not exchanged.

Extracurricular campus life included sports, drama, and music—

diversions that might have given young Skinner more social confidence. But here, too, he failed to achieve a real place for himself. He found that his tennis game, which had been passable in Susquehanna, could not compete at Hamilton. He auditioned for the Charlatans, a drama group, and got a part but was bumped in rehearsal. Having failed twice, he summoned the courage to try to play saxophone for an instrumental group and did find a niche there. Later he joined the Hamilton Glee Club. But these musical activities did not really boost his shaken self-confidence. Nor were his academic grades outstanding. With the exception of A's in algebra and trigonometry, he received B's in his courses during that first year.[22]

His favorite course was Paul Fancher's English composition. Fancher was a superb teacher who was rumored to be homosexual. A drama teacher as well, he read student papers in class with considerable effect. Despite his B in the course, Fred exhibited writing talent, and two of his poems were accepted by the *Hamilton Literary Magazine.* One had been written about the death of his Grandmother Burrhus in January 1923. The first and last stanzas of "Christmas Cactus" capture his sentimental and self-deprecating mood:

Oh, ugly loutish, selfish thing,
She cared for you
When you were naked, flowerless,
The whole year through

. . . .

She went, but you remembered all.
In her last hour
You bore to her, most gratefully
A blood-red flower.[23]

Fred's disillusionment and social isolation were best captured in a theme paper written in the third person in which he created a characterization of himself. Fancher had asked the class to describe the changes one year at college had wrought. In their senior year the themes would be returned to the students, who could then judge the accuracy of the earlier assessments. "In the Fall of 1922, a boy matriculated at Hamilton College," Fred began. "He was from a small town, reared in a sympathetic home, and trained in a school of interested teachers. . . . His home and school he thought had shown him the necessity for certain kinds of knowledge. . . . What a joyous task it was to be!" But, after eight months, "college had proved a disappointment." In fact,

> It needed barely one month of the first term to show the boy he had misjudged college. There was no majority of students who enjoyed study, who frequented the library voluntarily. He found that he was almost alone in his pursuit of literature, and that he was actually jeered at for spending time on a book when other boys were supporting athletics. . . . He wrote:
>
> "They're making me do too many things I don't want to do. They say these are things I need; yet, while they may know a lot about what the average person needs, they don't know half as much about *me* as I do."

The writer concluded that "the only broadening one year of Hamilton has given me is the enlargement of my own self-centered microcosm; the only agility of mind I have acquired is wasting itself in a ruinous flight toward selfishness."[24]

The cost of pursuing intellectual life and keeping his individuality was that "the boy . . . looked upon himself as isolated. He became critical, almost cynical. . . . The Great Change has been wrought." Yet it was not the one he had expected. And Fancher's comment on the back of his paper probably did little to uplift him: "There are three years more in which you may develop your individual bent. This is something encouraging to remember."[25]

Fancher had missed the point. Fred already had an "individual bent"—so much so that he was able to step outside himself and view himself as another person. His freshman year, with its social isolation and disregard for the importance of intellect, had caused him to imagine himself as another person, one who now stood apart from the anti-intellectual environment that surrounded him, one who had gained objectivity in the unhappiness of his social and intellectual predicament. This was the Great Change. Skinner was dispassionately analyzing his reaction to an unexpectedly punishing environment—not yet, of course, as a behavioral scientist but as someone newly detached from and cynical about himself as well as those around him. His suffering during that year moved Fred toward the detachment that characterizes the objective scientist.

IN THE SPRING a tragedy turned his attention to his family. Fred had concluded a singing tour with the Hamilton Glee Club and returned to Scranton for spring break. On April 7, while his parents were attending Sunday morning services at the Scranton Presbyterian Church, he, Ebbie, and a friend drove to a drugstore for sundaes and returned to the Skinner residence where Ebbie needed to use the bathroom. After a long while he emerged in great distress, saying he had an excruciating headache and needed to lie down. He asked for a doctor. One was called, but before he could arrive Ebbie fainted. Food was running freely from his mouth without the usual constrictions of vomiting. When the doctor arrived, he removed Ebbie's shoe and rubbed the sole of his foot. Fred rushed to church to get his parents, but they arrived home too late. Ebbie may have been dead even before Fred left. An autopsy performed that night showed he had suffered "acute indigestion," which "had caused an inflation of the heart [so] that the circulation of the blood had stopped completely with the heart attack."[26] Later, however, Skinner showed a report of the autopsy to a physician, who concluded that Ebbie had died at sixteen of a massive cerebral hemorrhage.

At the time of the tragedy, William and Grace were well established in an affluent section of Scranton. Their new residence, on North Washington Street in the elegant Green Ridge neighborhood, had been purchased for $14,500, a pricey sum in 1922. Grace had hired a maid and William had bought a Packard sedan, which then competed with Cadillac as the most prestigious American automobile. They belonged to the Scranton Country Club. Although they could afford the style of life required by the Scranton upper class, they still felt like nervous newcomers, acutely aware of their social inferiority.

Ebbie's loss devastated the Skinners, particularly William, and their lack of social connection in Scranton did not help to ease their pain. Ebbie had adjusted wonderfully to Scranton's Central High School, making many friends in a short time and earning a reputation as a good athlete. His proud father had recently bought him an Overland sedan, which the boys had driven to get sundaes that fateful morning.

For William, Ebbie's death brought a terrible disorientation: "The world was not the orderly, predictable thing it had seemed." The loss "haunted" him, leaving him lost and depressed.[27] He would later write a book on workmen's compensation law in Pennsylvania and dedicate it to "the Memory of My Son Edward." "The work," he explained, "was undertaken to afford distraction from the effects of his untimely passing."[28] For years after Ebbie's death, William would suddenly burst into tears when thinking of his youngest son.[29] With Ebbie's death, he had not only lost the apple of his eye; for Will Skinner, the American dream—the expectation that progress would come through the achievements of one's children, through the progression of the generations—was shaken to its roots.

Fred had watched his brother die with remarkable detachment. Just as he had dispassionately viewed himself in his freshman theme paper, so, too, he observed his brother's death as if he were a level-headed stranger happening upon the scene. He described the symptoms accurately and unemotionally, a fact not lost upon the attending physician. "With the same objectivity," Skinner recalled, "I had watched my parents as they reacted to the discovery that my brother was dead"—watched as his mother embraced the still warm corpse and his father walked in a trance exclaiming, "For heaven's sake, for heaven's sake."[30] Yet he was greatly moved by his parents' grief and the loss of his brother, remembering with mortification having accidentally wounded him with an arrow when they were children. Indeed, cool detachment and sentimentality were both characteristic of B. F. Skinner: "I tend to take major things of that kind [death] without any emotion, [yet] I think I am an emotional person. When something happens I accept it."[31]

Fred could accept Ebbie's death as fact, but it was harder to accept the new conditions that the tragedy created. He was now an only child, and this added measurably to his sense of being manipulated by others who did not share his interests: "With my brother's death, I was to be drawn back into the position of a family boy. It was a position I had never wanted, and it was to become increasingly troublesome in the years ahead."[32] Ebbie's death "threw my parents on me, searching for a boyish affection which he gave them but the demand for which was for me an embarrassment."[33] Yet he did nothing to resist being pulled back into their orbit: "[It] annoyed me, but curiously led to no open revolt."[34] Nonetheless, becoming the "family boy" left him feeling trapped. He had escaped to Hamilton College only to find that no one appreciated the student as intellectual; now, this sad event tied him closer than ever to his parents. He had wanted a new life but was in danger of lapsing into an old one. Fred was not only between two lives; he could not find a satisfactory life for himself in either place. By the end of his freshman year, having no close friends, male or female, his only pleasures were taking solitary walks and writing poetry. Returning to Scranton for the summer did nothing to help his mood: "I came back . . . a total stranger, and my parents had not yet come to know many people through whom I could make friends."[35]

Not surprisingly, William and Grace found life in Scranton less satisfying after Ebbie's death. The inevitable guilt a parent feels after losing a child was accompanied by second thoughts about having left Susquehanna; if only they had not moved, they pondered, he might still be alive. For some years Grace collected newspaper stories on the unexpected deaths of children, including many on the anguish of Charles and Anne Lindbergh over the kidnapping and death of their son, Charles.[36] Perhaps it helped to know that others, even the famous, suffered similar tragedy. They surely missed having old friends on hand who had known Ebbie and would have been able to console them.

The Skinners tried, perhaps too hard, to make Fred's first summer home from college enjoyable. He was introduced to the Hudson Coal Company's physician, Dr. John Fulton, whose daughter, Nell, shared his interest in music. The young couple often played duets, but their relationship remained strictly musical. William arranged for his son to play golf at the country club—hardly a social pleasure for Fred—and also allowed him use of the Packard to visit friends in Susquehanna. Like his parents, he felt ill at ease in Scranton: "we were upwardly mobile but with nowhere to go."[37] To escape their malaise as well as to ease their grief over Ebbie, they decided a seashore vacation would be therapeutic.

Nothing better revealed, however, the impossibility of recapturing the sense of well-being of times past. Stopping at Asbury Park, New Jersey, they found their accommodations crowded and the service indifferent. So they continued south to Spring Lake, and took rooms at the swank Essex and Sussex Hotel, which served the families of East coast bankers and brokers. Most of the guests were regulars who seemed a social notch or two above the Skinners. The class distinctions of the place, or at least their awareness of such, made Fred and his parents even more uncomfortable than in Scranton.

One incident at the Essex and Sussex well dramatizes their unease; not surprisingly, it involved Grace Skinner's old admonition to heed "what people will think." One evening they were served dinner in "a ritzy dining room with a snobbish head waiter with an Irish name." Fred remembered wearing "a buttoned sweater . . . and the head waiter said afterwards to my father, 'Please have the young man wear a coat in the dining room.' I can easily recall my mother's smothered cry of Uh-uh; we had all made a mistake."[38] What had been envisioned as a relaxing interlude became a vacation pervaded with misgiving and social discomfort. And Fred suffered the double disapproval of both the hotel employees and his mother.

The result was vacationer paralysis. Fred was unable to muster the courage to approach unattached girls, and he remembered envying a young man with a sports car who casually tipped the doorman. The alternative to making new friends was being with his parents, so he took frequent solitary walks. He had plenty of time to speculate about his future, and it did not look promising. His prospects at Hamilton appeared to be a continuation of isolation and disappointment; and Ebbie's death meant he faced spending more time with his parents. "I was as miserable as I have ever been," he remembered. "My parents were scarcely less so."[39]

He could not, however, openly confront them with his own unhappiness, especially in their time of sorrow. Besides, Fred was not a young man who could easily tell his parents they had silly class pretensions, because to some extent he shared them. The Essex and Sussex sojourn accentuated their common maladjustment after leaving Susquehanna, their isolation, their inability to measure up in unfamiliar surroundings. And although Fred's discomfort was not quite the same as that of his parents, he remained remarkably sensitive to social distinctions. His mother and father wanted the social standing that accompanied affluence; all three fervently wished to avoid the stigma of social inferiority.

BUT WHEN FRED returned to Hamilton in September, good social fortune unexpectedly came his way through a close association with one of the college's most intellectually accomplished and artistically cultivated families—a circle in which he soon felt entirely comfortable. Still between two lives, he began to feel the gravitational pull of an alternative family culture. His social isolation and absence of intellectual companionship ended.

Arthur Percy Saunders served as both college dean and chemistry professor. His family's home was a salon, drawing a continual stream of well-known literati and artists as well as a select group of Hamilton students to its culturally sophisticated yet relaxed atmosphere. Ezra Pound, Robert Frost, Alexander Woollcott, Ivor Armstrong Richards, and James Agee knew the charming conviviality of the Saunders's music room. There the professor himself played the violin and held regular concerts, mostly violin quartets and piano recitals.

Described as "a rare man in a hurly-burly world, as a man of science, as a man of grace, charm and urbanity, whose hospitality was the same at home and abroad," Saunders was also a "man whose sense of values has impressed and directed the lives of hundreds of students."[40] He was an outstanding gardener who raised prize-winning hybrid peonies, became president of the American Peony Society, and left records on some seventeen thousand plants. Though not a professional writer, Saunders loved literature. He was also an amateur astronomer who watched the starry heavens on summer nights through a telescope set up in his garden. A political liberal, he subscribed to radical magazines such as *Broom*. His liberal leanings disposed him to look unfavorably on Hamilton's president. He found Ferry's pro-business outlook, his manipulation of the faculty, and his lack of due process in disciplinary cases with students arbitrary and autocratic.[41]

Saunders shared his house, just west of the Hamilton campus, with his wife, Louise Shefield Brownell Saunders, and their children. One of the first graduates of Bryn Mawr, Louise was a cultivated woman who in 1897 was appointed warden of Sage College, the women's counterpart to Cornell. She also lectured on English literature and met Percy while he was teaching chemistry at Cornell. She resigned her position when the Cornell administration refused to name her a professor. In Clinton, she tutored Hamilton students. She and Grace Root, the wife of the art professor at Hamilton, were known as "the most formidable and influential women of the Hill for something like three decades."[42] There were four Saunders children: Silvia, an accomplished singer; Olivia, known as "Via," whom Fred took to the junior prom and who later married James Agee; William Duncan, already a talented poet in his teens; and Percy Blake, nicknamed "Frisk," who loved natural history.

Tragedy had struck the Saunders family in January 1922 as it would strike the Skinner family in April 1923, with the sudden death of eighteen-year-old Duncan. A freshman at Hamilton, he had been tussling with some fraternity brothers when his head violently struck a wall. Soon he began to suffer nausea and incoherence. He died about twenty-four hours later, after undergoing emergency surgery in a Utica hospital to repair brain damage. William Duncan had been a promising young man, the family favorite, and after his death his grief-stricken parents and siblings decided to take an extended European trip. This coincided with Fred's freshman year, so it was not until his sophomore year that he made contact with the family that would so dramatically improve his life at Hamilton.

Fred, about the same age as Percy and Louise's deceased son, was chosen to tutor "Frisk," their youngest, in mathematics. This arrangement not only put Fred on friendly terms with Hamilton's most intellectually cultivated family, but through it he also came to know the Roots, for generations the college's most socially powerful family. The most famous Root, Elihu, was chairman of the college's Board of Trustees. He had served as secretary of war for presidents William McKinley and Theodore Roosevelt as well as secretary of state for the latter, and was often mentioned as a potential presidential candidate himself. Indeed, it was one of William Skinner's oft-repeated and incorrect predictions that Root would someday hold the nation's highest office. An invitation to the Root home was considered a great honor. For his part, Fred thought Grace Root, Elihu's daughter-in-law, regal and wise, and she took a special interest in him.[43] The Root home lay adjacent to the campus and the family owned land known as Root Woods, where Hamilton students, faculty, and lovers often strolled.

Fred found the style of life at the Saunders household pleasantly different from that of his parents. For one thing, they had a more relaxed moral code, not libertine but certainly not Victorian. Percy had an eye for the young ladies, and would eventually carry on an affair with one of his daughter's friends. The girl, under the care of a New York City psychiatrist, told her doctor of the liaison, who then, unprofessionally, passed the confidence on to the girl's parents. They confronted Louise, but she refused to be shocked or embarrassed. Instead she went to New York, met the psychiatrist, and defiantly defended her husband.[44] Such a disclosure, let alone defense, would have been unthinkable to Grace Skinner.

The Saunders's house was an intellectual and cultural mecca where pursuits from astronomy to chamber music and horticulture were encouraged. Like Mary Graves, this family took Fred under its wing, but this time intellectual life was combined with social stimulation. It was an invigorating family culture, a house adorned with literature purchased with care from bookstores rather than bought on a whim from traveling book salesmen. The Saunderses offered Fred a novel family culture without the moral policing he had known at home. Fred had just enough pretentiousness as well as genuine intellectual curiosity to be susceptible to this kind of environment.

Percy Saunders encouraged the nonconventional yearnings of youth, particularly the ambitions of a young man who was an outsider and a bit of a rebel. As he explained it:

> If the young show radical tendencies, i.e., show themselves not perfectly satisfied with the status quo—which gives [the older students] wealth in the business world and power in the educational—they are thrust back and kept under as far as possible.
>
> But if a young man shows himself content with things as they are . . . he gradually becomes known as a "safe" man and to him power is more and more given, for those higher up are satisfied he will do nothing to disturb the established order. Once in a while a radical, one who does his own thinking, instead of putting it out as we say of the wash—comes to the top. Then he is in trouble.[45]

Saunders appreciated young men who did their "own thinking"; William Skinner liked the "safe" men. While Fred believed that his father's faith "in progress may have had a stronger effect on me than I have realized," it was through contact with the Saunders scene that he realized "there was a better world . . . and learn[ed] how to behave in [it]."[46]

Fred's parents had built their world in accordance with the canons of the Protestant ethic, complying with its codes of social mobility and material success. Who you were and what you had was a consequence not only of the previous generation's status in life, but of your own efforts. In the 1920s a vigorous boosterism epitomized by the adage "to think success brings success" had gained enormous leverage in America as the advertising techniques of Madison Avenue gained ascendancy.[47] Will Skinner was active in the Scranton Kiwanis Club, and his 1928 election as its president showed that he well represented the post–World War I business creed. While the Saunders family certainly didn't reject affluence and, in fact, lived quite well, they loathed marketplace values. They were not original in this, following the example of literati who, although willing to be subsidized by the American market, considered themselves apart from the business class. They echoed those enclaves of departed American intellectuals and artists who flowered in mid-nineteenth-century New England and who created an American Renaissance: Emerson, Thoreau, Hawthorne, Melville, Whitman. At the center of their shared creed was reverence for the individual genius whose literature or art transcended the mundane and the material and represented universal truth and beauty.[48]

Finally, the Saunders household was a lively one, where interesting people not only came and went but also lived. Percy and Louise often boarded promising young people whose parents wanted them to be tutored and socially prepared for college. There was, no doubt, a certain snob appeal in catering to well-heeled families who wanted their sons and daughters to acquire the cultivated tastes of the upper class. Nonetheless, these protégés experienced a way of life that emphasized intellectual excellence and aesthetic standards—an environment that enthralled Fred. Perhaps the greatest attraction was that in the company of the Saunderses one felt *both* intellectually and socially accomplished—even superior. Cynthia Ann Miller, daughter of a Utica banker, was one such boarding student, being groomed for Radcliffe College. Fred met her while he was tutoring young Frisk Saunders and promptly fell passionately in love.

His high school infatuations had left him feeling ignorant, inept, or guilty. He had tended to approach girls as either potential sexual conquests or idealized romances. His sophomore year crush on Cynthia Ann Miller fell into the latter category: "My love for Cynthia Ann was deep and painful but it was not primarily sexual. Indeed, I must have seemed sexually backward."[49]

The Saunders scene contributed to the intensity of his new romance. The couple had tea in the music room; they walked hand in hand in the surrounding woods; they visited Grace Root; and they read and discussed poetry, some composed by Cynthia Ann. All the Saunderses knew that Fred was enamored, and when Percy and Blake learned that Cynthia Ann was also being courted by an older student, they could not resist teasing Fred by singing the then popular song "Somebody Stole My Gal." It was not just the family who knew Fred was smitten; the whole campus seemed aware of it. The college yearbook, *The Hamiltonian,* reported: "Fred claims that he is not sentimental but sophomore year he was caught on a moonlit night leaning out of a window in South College while inhaling the aroma of a violet-scented vanity case to the pathetic yearning of a broad 'Ah!' "[50]

One day in early 1924 while he was walking through Root Woods, it became clear to Fred that Cynthia Ann wanted to make love: "We partially undressed. . . . I don't think I knew the first thing about what was expected. Anyway, I didn't go through with it. It wasn't long after that that she wrote me this long letter breaking it off."[51] Fred was crushed and "went for months in an agony of unrequited love. I watched for her everywhere."[52] Others, however, seemed to know the relationship was doomed. Olivia Saunders recalled that it had been a one-sided romance.[53] Skinner kept in touch with Cynthia Ann after she went to Radcliffe and after she married, but he suspected that she could have found a more suitable husband—someone, perhaps, more like himself.[54]

His unsuccessful courtship influenced Fred's literary interests. His sophomore year he took a course on French drama and especially liked Edmond Rostand's poetic rendition of *Cyrano de Bergerac,* not only because of his interest in French literature but because like de Bergerac, he, too, suffered from unrequited love. But romantic disappointment did not drive Fred into the miserable isolation of his freshman year. Being a part of the Saunders's circle improved his standing—indeed, the failed courtship and the family teasing drew him even closer to the family. And his new associations encouraged others. Fred became a regular customer at Mary Ogden's book shop in Utica, an establishment frequented by Percy and Louise. There he befriended an attractive clerk who introduced him to the works of Freud and talked frankly about Victorian repressions. The bookstore was another place where he could discuss subjects that were taboo in the Skinner home.

Fred's sophomore year was the pivotal one in his college career, at least in how he felt about being a student and in terms of an emerging academic interest. Embarrassment about slips in class diminished as he made gains socially and intellectually. He was enthusiastic about most of his courses, which, along with French drama, included declamation, debate, English composition (again with Fancher), Greek (a study of Homer), general introduction to English literature, psychology (that is, logic), and elementary chemistry.[55] Although he did not care much for chemistry, he loved his writing classes and still hoped to be a writer.[56]

He also found a best friend, one who shared his intellectual interests and writing ambitions. John ("Hutch") Hutchens, also a sophomore, had arrived at Hamilton College from Montana. He had also experienced an unhappy freshman year, nearly flunking out. But, unlike Fred, he had pledged one of the most prestigious fraternities on campus. Hutch was more outgoing than Skinner, and was recognized as "the lad with the most angelic face and diabolic mind in Hamilton College."[57]

The Hamiltonian mocked Hutchens as living "under the terrible hallucination that he is the reincarnation of Shakespeare, Milton, Dante, Molière and company."[58] This literary reputation was enhanced by Hutch's contributions to "Carpe Diem," a satiric column in the college newspaper, *Hamilton Life.* There he poked fun at Fred's romance with Cynthia Ann. Writing under the pen name John Kay, he composed a Chaucerianlike poem entitled "Sir Burrhus Goeth Forth," which pilloried that bold knight's failed attempt to capture the love of "Ladie Fariana." Sir Burrhus, however, took up the challenge and published an equally humorous rejoinder.

Together they pilloried conventional standards, especially the pretentious, materialistic values of some Hamilton professors and college administrators. In these attitudes and activities they were similar to other intellectually inclined American students in the 1920s who maintained a "smart-set" persona, a cynical post–World War I disillusionment with Victorian traditions and contemporary business-oriented values. Aside from thawing Fred's reserved manner by introducing him to the satiric fun of writing for student publications, Hutch accompanied him on their frequent Saturday-night forays into Utica for dinner, drinking, and "investigat[ing] certain educational aspects of life not specified in the college catalogue."[59]

IN FRED'S THIRD ACADEMIC YEAR at Hamilton, he took junior declamation, discussion, debate, elementary Spanish, English literature (seventeenth and eighteenth centuries), American literature (contemporary), art appreciation, American government, and anatomy and embryology.[60]

It was in anatomy and embryology that Fred returned to the hands-on activity he had loved in Susquehanna. In this class—the only science course at Hamilton that related to Skinner's career as a behaviorist—he recalled dissecting a cat and making slides of chick and pig embryo. Professor Albro "Bugsy" Morrill recommended that he read the early-twentieth-century biologist Jacques Loeb's *Physiology of the Brain and Comparative Psychology* (1912) and *The Organism as a Whole* (1916), and Loeb's concept of tropism greatly impressed him. Tropism is movement that can be controlled by the way a scientist exposes a simple organism to light. The control of the behavior of organisms would be essential to Skinnerian science.

Professor Calvin "Cal" Lewis taught a public-speaking course about which Fred wrote a satiric piece entitled "The Confessions of a Puzzle Eater," the story of a crossword puzzle addict. Lewis, Hamilton's great authority on creative writing who also offered a course on the novel, was so incredulous about the authenticity of Skinner's essay that he accused him of plagiarism—a gross violation of Hamilton's honor system, which, if confirmed, could have resulted in expulsion. But Lewis later apologized; he had not wanted to admit that a Hamilton student could write so professionally.[61] The incident boosted Fred's emerging conviction that he should make a career as a writer.

His literary ambition was strengthened even more during the summer of 1925, following his junior year, when he attended the Bread Loaf School of English in Vermont. There, deep in the Green Mountains on a 30,000-acre property, students had the opportunity to attend a summer school led by nationally renowned writers and dramatists. That summer two American legends, Carl Sandburg and Robert Frost, were in attendance. Frost had recently received the Pulitzer Prize in literature and was, like Skinner, at Bread Loaf for the first time. Frost suggested that Skinner send him samples of his writing, a gesture that must have made his spirits soar as well as causing him considerable anxiety.

While at Bread Loaf, Fred also fell in love, this time with "Ellen" (as he called her in his autobiography), a married woman in her late twenties who was accompanied by her four-year-old boy. Skinner followed her like a lovesick puppy but, although they took walks together and had an incipient romance, she rebuffed his sexual advances. In a writing class taught by Sidney Cox, attended also by Ellen, Fred wrote a disguised account of the friendship in which he alluded to his love's resemblance to his mother.[62] Ellen understood the implication immediately and gently kidded Fred after class about falling in love with a woman who resembled his mother. Once again he experienced the pain of unrequited love and blamed himself: His love was being rebuffed because he did not know how to make love; sexual naïveté was a Susquehanna legacy that his experience with Cynthia Ann and now with Ellen had not overcome. Both courtships augmented his feelings of inadequacy.

There were, however, other, if less romantic, solutions to his problem.

Hutch was more sexually experienced than Fred, having frequented houses of ill repute in his native Montana. During their senior year the two friends, along with Jack Chase, the Latin professor's son, made the first of several trips to Utica's red-light district. Hutchens believed the escapades had a valuable long-term effect: "I can think of a dozen men I've known who would have been saved from early, foolish marriages if they'd received release in a good, well-run whorehouse instead of marrying the first girl they went to bed with."[63] Fred, on the other hand, once again met disappointment and disillusionment. But, thanks to Hutchens, he was at least gaining sexual experience, if not romantic fulfillment.

SKINNER'S FINAL YEAR at Hamilton made him more experienced in academic ways as well. His course work further emphasized his literary interests: senior oration and debate, advanced grammar and modern prose, English language (Anglo-Saxon and Middle English), English composition, Shakespeare and Elizabethan drama, and advanced art.[64] Fred became assistant editor to Hutchens's editorship of the *Royal Gaboon,* the most irreverent of the campus publications. Under their purview it became more intellectual, merging with the *Hamilton Literary Magazine.* Fred contributed book reviews and wrote an article on Ezra Pound—"A Great Hamilton Alumnus, Unknown to Hamilton Men." As "a romantic figure," Fred wrote, Pound was "quite naturally . . . misunderstood by the student body" when an undergraduate; but that misunderstanding could now be corrected if Hamilton students were to discover the "pleasure in acknowledging a great man to be one of them." Fred concluded: "We, who of all others should find a 'kinspirit' in Pound, ignore him."[65]

Did Skinner identify with Pound as a "kinspirit" intellectual who, like himself, had been ignored by his classmates but would go on to literary greatness? There were certainly other writers, like Frost, whom he admired more at the time. But here, as a measure of his changed status, he was writing not for himself and his professors but for the student body—intellectually, as it were, hazing them.

Fred and Hutch often vocalized their literary association across campus, as one would shout a prolonged ROY-AL GA-BOO-N and the other would echo it. Their literary interests took a competitive turn when as seniors they were both up for the two-hundred-dollar William Duncan Saunders Prize for creative writing. To ease any ill feeling should one of them win, they agreed that the winner would give the loser seventy-five dollars. "I was not shaken," Skinner recalled, when Hutchens handed over the consolation prize. "There was no doubt that Hutch was good."[66] The short story Fred submitted, "Elsa," developed a theme that would long remain with him. Told from the fictional Elsa's point of view, it is the story of a marriage in which her powerful but foolishly idealistic husband subtly forces her to bend to his will. Rather than separate, they stay together under unhappy, though not unbearable, conditions.[67] Fred, of course, did not face an aversive spouse; he faced aversive parents and wanted to break from the control of their way of life. He later concluded, "I had not broken on little things and therefore had not discovered *how* to break."[68]

But he had discovered how to make trouble covertly for those at Hamilton who posed as something they were not. One episode illustrates well this covered rebellion and smart-set cynicism. He and Hutchens took aim at Fancher, the English composition professor, whose nickname, "Smut," referred more to mannerisms than to a love of cheap literature; he had a reputation for being a notorious gossip and name dropper. He gave the impression, for example, that he knew everyone who was anyone in theater and film. During the fall term of their senior year, the two friends schemed to make it appear that Fancher had arranged to bring Charlie Chaplin to Hamilton for a special lecture entitled "Moving Pictures as a Career." Fred arranged for a high school friend working at a nearby newspaper to print bright orange posters advertising Chaplin's appearance, with a precise date and time.[69] In the early morning hours of the scheduled day, Fred and Hutch tacked up posters all over Clinton. The Utica newspaper even announced that Hamilton's president would make a statement about the event at chapel. Phone calls began to pour in, but, of course, neither the president nor Professor Fancher knew anything about the concocted lecture.

The college administrators discovered the hoax by noon and tried to avoid public anger and disappointment by arranging for the local police to meet incoming cars and inform visitors of the chicanery. Nonetheless, several hundred got through and, upon seeing students gathered for a pep rally, assumed Chaplin was on campus. The next day the two conspirators, still in the heat of their sarcasm, penned an editorial for *Hamilton Life* arguing that "no man with the slightest regard for his alma mater" would have perpetrated such a deceit.[70] The college administration undertook an investigation and even hired detectives to track down the culprits.

What began as a practical joke to alleviate boredom and strike out against pretense had turned into a real fear of being expelled. Fred confided his complicity to Percy Saunders, who advised him to keep a low profile. For days he fretted that, since he had used his own typewriter for some of the print on the posters, the authorities would trace it back to him; he filed off the corners of several keys. The police, concentrating on Jack Chase, a known prankster, never pinned the blame on anyone.

IN JANUARY 1926 Fred began to make plans for his life after graduation. He wrote his parents of his plan to live with them for one year and try his hand at writing a novel. They agreed to his proposal but with little enthusiasm and strong reservations. William wrote back to him: "You will find that the world is not standing with outstretched arms to greet you just because you are emerging from a college." He reminded Fred of something he had learned from the summer at Bread Loaf: Even exceptionally talented writers often have financial difficulties. It was perfectly acceptable to try to become a writer, William wrote, but "I don't want you to become one of those hermits who live in a garret on a crust of bread . . . and . . . will not condescend to get down to earth and mingle with the common trash or be as others are."[71]

Will Skinner, fearing for his son's financial future, did not share Percy Saunders's assumption that intellectuals are superior to businessmen. Appreciating his son's well-rounded education, William nonetheless resented his untraditional intellectual ambition. He wanted Fred to be more successful than he, but through the usual American avenues of upward mobility: business and the professions.

His uneasiness notwithstanding, William mixed praise with concern. He explained to Fred that "I believe you have the ability to study and master about anything you care to undertake," but also considered the possibly negative outcome: "Suppose for instance that your dream does not come true. Are you going to be disappointed and feel sour and enter other lines with lack of interest and distaste? . . . [And] how are you going to account to your friends for a year's apparent idleness and the impression that would give them that you were lazy?"[72]

William, too, was influenced by "what people will think." After all, he and Grace would have to explain to friends why their son, a college graduate, was unemployed while country club friends proudly announced that their sons had landed respectable positions as accountants and engineers or were going on to law or medical school. He urged Fred not to dismiss his advice just "because it is made up of old-fashioned platitudes, or because it reads like 'Letters from a Self-made Merchant to his Son.' " Fred's long-term success was his essential concern: "if your talents enable you to do something big and startle the world no one of course would rejoice more than your mother and I who have our whole life centered in you and your success."[73] How Fred must have dreaded reading this sentence. He was still very much under the parental thumb; indeed, they seemed more involved than ever in shaping his future.

He took his father's letter to Percy Saunders, who did not, as he expected, wholly condemn it. Saunders was skeptical about the genuineness of Fred's goal and afraid Fred would find himself trapped by it.[74] With Saunders's reservation and only tepid support from his parents, Fred's plans for a writing career appeared uncertain.

Then, less than two months before graduation, he received the confirmation he sought. Robert Frost had read the short stories he had sent after leaving Bread Loaf. "I ought to say you have the touch of art," the poet judged: "You are worth more than anyone else I have seen in prose this year." This was indeed high praise, and, although Frost could not guarantee Fred would be a successful writer, he was certain what made one:

> All that makes a writer is the ability to write strongly and directly from some unaccountable and almost invincible personal prejudice like Stevenson's in favor of all being happy as kings no matter if consumptive, or Hardy's against God for the blunder of sex, or Sinclair Lewis' against small American towns, or Shakespeare's mixed, at once against and in favor of life itself. I take it that everybody has a prejudice and spends some time feeling for it to speak and write from. But most people end as they begin by acting out the prejudices of other people.[75]

Ecstatic, Fred took the letter to Saunders, who now agreed he should try making a career of writing. The letter had made it much easier for Fred to accept his parents' weak support. That Frost's own literary success was hard-earned and a long time coming was irrelevant; his judgment not only confirmed Fred's opinion of himself but played to his conceit. By sharing the news with his parents and Percy Saunders, he removed objections and further congratulated himself on knowing himself best.

WHEN WILLIAM AND GRACE SKINNER came to Hamilton to attend their son's graduation, the Saunders family invited them to tea. They had often entertained the conventional parents of their live-in protégés and made a valiant but failed effort to put the Skinners at ease. But Grace's voice "tightened up," and William "tried to say the things he supposed appropriate."[76] The most uncomfortable of all was Fred, for his two separate worlds lay before him in the same room: "I had developed two verbal repertoires, appropriate to very different audiences, and now the two audiences had come together and there was little I could say that was appropriate to both."[77] (Skinner's book *Verbal Behavior* [1957] would deal with the crucial role an audience has as a controlling variable in shaping one's verbal behavior.) But his discomfort at the graduation tea reached beyond different "verbal repertoires" to the uncomfortable social position of being in the presence of two different families who had shaped him in profoundly different and conflicting ways. What may have made Fred the most uncomfortable of all was the realization of how closely tied he remained to Grace and William, even though he had largely rejected their culture.

At the graduation two student addresses were given. The valedictorian spoke on "Plymouth Rock and Ellis Island in American Life." Fred Skinner, salutatorian once again, was required to give his address in Latin. After parodying college trustees as "owners of shiny automobiles" and President Ferry as "sweet talking but most vehement in action" (was he alluding to Ferry's mighty efforts to find the perpetrators of the Chaplin hoax?) he praised the professors for the results of their four years of labor. To his classmates, he was kind: "whether eager students or slackers, bookworms or athletes . . . 'we have seized the day.' " He ended by making fun of himself: "My wit languishes, afflicted by long use."[78]

Skinner had used this address to poke fun at those who represented the conventional college culture, or those who had caused him suffering. It showed he had learned something at Hamilton besides how to cheer the football team and make good grades. Being an intellectual meant being honest about those around you, even if to do so you had to use Latin satirically. But the satire did not reflect a lack of respect for Hamilton

College; for decades, well into his eighties, Skinner often showed up for the class of '26 reunions.

Years later, Fred believed that he left Hamilton a different man than he had been upon his arrival or than he expected he might be at graduation: "I think that the classical education I got at Hamilton was very important," he said. "I had the courses that make one an intellectual."[79] In addition, his association with the Saunders family had made him a hybrid, like one of Percy Saunders's carefully grafted peonies. By remaining attached to his family after he had experienced the intellectual and moral freedom epitomized by the Saunders family, Fred achieved a tenuous compromise, one he could not suspect would bring a year of personal unhappiness.

3

A Hill of Dreams

I am reminded of Arthur Machen's Hill of Dreams, *which I must have read nearly sixty years ago. I have forgotten most of the book but I remember this. The hero is a writer who thinks he is creating masterpieces. After his death they find hordes of empty little blue bottles. He has been on drugs.*

—B. F. Skinner, Basement Archives, 1986

In June 1926, Fred returned to his parents' three-story frame house in Scranton. The affluent Green Ridge section was a pleasant neighborhood of tree-lined streets and imposing homes, nestled at the bottom of one of the city's many sizable hills. The Skinner residence was situated a block or two from the more exclusive homes. The maid lived in a room on the third floor. Between it and an attic storage area was a room with a sunny southern exposure. Here Fred tried to become a writer during what he later called his "Dark Year"—actually eighteen months.[1]

Fred's first creative work in Scranton was not to scribble the opening sentences of the great American novel or even a short story but to fashion another box of sorts, a suitable writing place. He built himself a bookcase and a work table. He bought a filing cabinet for the many manuscripts he expected to produce. And he constructed a rack that could hold a book in a convenient position across the arms of a chair, where he read the novels of Sinclair Lewis, Fyodor Dostoevsky, Marcel Proust, and H. G. Wells, as well as contemporary literary journals such as the *Saturday Review of Literature* and *American Mercury* and little magazines such as *The Dial, New Masses, Two Worlds Monthly,* and Ezra Pound's short-lived *Exile.*[2] "I have constructed for myself a study," he wrote Percy Saunders. "Here I can retreat when the ghouls of conservatism become too annoying and read the *New Masses.*"[3]

Sometime during this period Skinner recalled reading Arthur Machen's novel *Hill of Dreams.* Skinner's memory of the novel, quoted at the start of the chapter, was a bit faulty. It depicted a young man's unsuccessful effort to write, but his suicide resulted from the agony of being a failed writer rather than a drug-induced writer's euphoria. Like Machen's character, Fred shared a growing alienation from his parents and the conventional life around him as he attempted to write in an atmosphere of intellectual isolation. About his lack of success he would later conclude: "I had failed as a writer because I had nothing important to say."[4] He might have added that the collapse of his high expectations, especially after Frost's encouragement, was a bitter disappointment.

During the period he devoted to writing, Fred developed a growing curiosity about writers who embraced a behavioristic philosophy of science. His correspondence with Percy Saunders shows how his confidence in a writing career dissolved into another form of intellectual isolation, one that brought him to a point of view that could be called behaviorist even though he had done no scientific work in the field. Years later, looking back on this period of his life, Skinner said he was both a writer and a behavioral scientist at that time. "I did write a few fairly good things," he explained, "but when it came time to do nothing but write and to make my name as a writer, I failed miserably. . . . Fortunately, I was almost accidentally able to acquire a different repertoire which worked much better and had happier results."[5]

Skinner recalled that the clinical psychologist Henry Murray once called him "a romantic defending himself with science."[6] Fred's letters to Saunders suggest that Murray hit upon a way of squaring Skinner's emotional yearnings with his objective science. During this period Fred had what might be profitably called a romantic crisis. He was struggling to protect himself against the implications of certain experiences he could not satisfactorily understand by appealing to religion, philosophy, or any traditional intellectual rationale. His crisis was romantic in the sense of being a young man's lonely, even heroic battle against conventional values; in his self-conscious attempt to become a writer/intellectual; and in his personal longing for a future he could not yet realistically envision. The shifting moods, the hopes, disappointments, and fears of his Dark Year—a strikingly romantic label—were accompanied by his conversion to behaviorism as a philosophy of science, one that denied that consciousness could be studied objectively through introspecting what was in the mind. Skinner embraced behaviorism not as a cold, unfeeling nihilist but as a sensitive, unhappy, slightly cynical young man, who in fact was searching—at times desperately—for something not to be cynical about. He would remain a romantic all his life.*

"This letter isn't important," Fred disclaimed to Saunders, in mid-August 1926. "Put off reading it until some evening when a cigar and fire in the music room will seem agreeable." Far from being enthusiastically engaged in his writing, he was thoroughly discouraged—and he had been back in Scranton barely two months. "The main thing about this letter is my writing it; it's a good dose of castor oil to me for I'm dying from a congestion of ideas."[8] The "congestion of ideas" was a blockage of his ability to write literature he considered original, although he did attempt some short stories.

What had gone wrong? "For one thing," he admitted, "I have been terribly depressed for over a month now, so depressed that I haven't written a word." Obviously his depression and inability to write were interrelated. Yet he did not say that he was depressed because he could not write. Rather, the depression had emerged for other reasons, one of which concerned the *nature* of his writing. "The only kind of writing which fits my idea of pure literature," he wrote to Saunders,

> is objective writing. I can't honestly or dishonestly do any other kind. But look what one meets up with in writing objectively: 1) not one person in a hundred understands you—not me alone but any objective writer—the populace needs interpreters; 2) those who do understand you won't give you credit for doing the thinking but take it all to themselves. . . . 3) the volume of objective writing necessary to express a philosophy of life.[9]

Objective writing was exclusively descriptive writing; the writer did not talk about the feelings of his characters or put thoughts in their heads.

*I don't mean explicitly to associate Skinner's romanticism with, for instance, nineteenth-century German romanticism, or with any other European or American romantic movement. But David A. Hollinger has recently analyzed an interesting division between "knowers" and "artificers" among post–World War I modernists. The former tended to be, among other things, concerned with "demystifying," while the latter were, among other things, more "myth-constructing"; more broadly, this was a distinction between the scientist-knower and the artist-hero. During the Dark Year Fred struggled as if he were the latter, but he moved toward the former with his failure to become a writer. In this sense, Skinner saved or defended himself from romantic, or artistic, failure by discovering behaviorism.[7]

Nor was a clever plot essential to a successful story. The reader discerned the feelings, thoughts, and story line. Skinner remembered being intrigued when a bright girl he had known in Susquehanna quoted G. K. Chesterton on a character of Thackeray's: "Thackeray didn't know it but she drank." And he remarked: "A writer might portray human behavior accurately, but he did not therefore understand it. I was to remain interested in human behavior but the literary method had failed me."[10]

Skinner's objective writing was similar in spirit to his "descriptive behaviorism," as it was called early in his career. But it did not seem original in the sense that the objective writing led him to new discoveries. Skinner wanted to know that what he wrote was accurately descriptive. To fail to know the characters one created was unacceptable. When his former professor of composition, Paul Fancher, had observed that even Fred himself did not understand what he had written—a point virtually the same as Chesterton's—Skinner knew that true objective writing was perhaps impossible. Even so, he might have discounted this difficulty if he had found someone who could have taken the role of sensitive literary critic. Perhaps that is what he hoped for from Saunders.

There was also the problem of producing enough literature to express a philosophy of life. He wanted to have his writing make a statement, announce some principle or truth that lay beyond objective writing but at the same time encompass it. In a note to himself he criticized Anton Chekhov's short stories because they expressed no philosophy of life, while praising Dostoevsky's *The Brothers Karamazov* for having one—although Dostoevsky was not an objective writer. His frustration and depression stemmed from the incongruity that "the literature I was philosophizing about in this way could scarcely have been further from the literature I was producing."[11] Skinner wanted objective writing but gradually realized that there was no such thing.

OTHER CONCERNS also weighed heavily upon him during the first months of the Dark Year. Two experiences in particular had an enormous influence on him, both of which he recounted in a letter to Saunders.

Shortly before Fred's graduation, seventy-seven-year-old Grandfather Burrhus had undergone surgery for an enlarged prostate. He seemed to recover normally but then fluid collected into a hydrocele, necessitating another operation from which, once again, he quickly rallied. He went from the hospital to the house on North Washington Street, where he allowed no one but Fred to tend his dressings and clean the fistula through which his urine drained.

After a few days he died of bronchial pneumonia, a death Fred observed at a remove:

> For a day and a half I watched him—he was apparently awake yet unconscious save for neuralgic pains—he hiccoughed badly—until the last evening they gave him morphia. This depressed his lungs which brought on coughing and his right lung filled completely within an hour. Then all night long this organism—worn out, beyond repair, lay there. Certain muscles of his diaphragm went on functioning—a little air was pulled spasmodically into the remaining lung space. An overtaxed heart—sustained on strychnine—pumped impure blood—and gave out under the strain. His pulse weakened—he coughed a bit and lay still. I listened to his heart—it was still. I lifted him up—a little black fluid ran from his lips.

He watched unemotionally as "this organism" expired, and tried to understand it:

> What had happened? The active *idea* which I had known as my grandfather was gone simply because certain physical properties of his body had given out. Was there anything more of him beside that, something spiritual? If so, when did it leave him? At the last moment?—except for certain reflex muscular activities the minute before and minute after [death] were alike.
>
> I am very sure that my grandfather—all of him—all that I knew of him and felt—his character, personality, emotions, skill, desires—all—everything went as soon as the physical condition of his body became unfit for certain nervous coordinations. Just as the dreary character of the clock I now hear will vanish when the parts which give forth its ticking shall stop.[12]

Just as he had avoided giving feelings and thoughts to his fictional characters, Fred did not record subjective states such as how his grandfather must have felt or looked. His description consisted of observable physical reactions. He knew, though, that his grandfather's death meant more than the cessation of physical functioning; there was an ontological consequence of *being nothing* beyond the termination of reflex action. Fred asked "what had happened?" without the traditional supports of religion and metaphysics. His focus on observables and his reluctance to go beyond them would be characteristic of his scientific approach in the 1930s. All that was real, all that could objectively be said, had been said in Fred's description of the physical behavior of the death of an organism that a moment before had been his grandfather. It was probably his first written description of the behavior of a whole organism and was similar in tone to the way he later recalled his detachment in describing Ebbie's death to the attending physician. Objective detachment and a gift for terse, accurate description had forcibly impressed upon him that there might be nothing more to an organism than its behavior.

Fred's clinical description of his grandfather's death did not mean that he did not feel sorrow and emotional pain any more than it meant he had not felt badly for Ebbie and his grief-stricken parents. Some might conclude that Skinner's imaginative powers had failed him; but he was making a crucial distinction between guessing about the subjective, inner life and making a simple description of observable phenomena. Similarly, the behavioral science he would develop never denied that feelings existed, but it did not appeal to the unseen for its facts.

A second shaping experience that Fred underwent in those first months out of college was arranged by Dr. John Fulton, the physician who had operated on his grandfather. Fulton attended the same church as William and Grace but, like Fred, only went when he was "dragged."[13] Fred enjoyed Fulton's dramatic recitation of poetry. Seeing Fred's interest in writing falter, Fulton encouraged him to consider medicine and suggested that he observe some operations. Again with cool detachment, Fred described to Saunders a surgical procedure that emphasized the thin, fragile line between life and death:

> I put on a white gown—white cap—stand beside an operating table—watch my friend the doctor operate on a broken back. Ether—a breathing body under a white cover—a square hole in the white showing iodine-painted skin. A long slow cut—a rolling out of blood—vertebrae exposed, chewed off with forceps—pieces of ivory-like bone crunched out—three inches of spinal cord exposed—mangled cord. Here is an ounce of tissue—crushed—meaning a life of complete paralysis—or, better, death. How far apart are life and death? One uncontrolled pressure from the doctor's forceps—and instant death. Controlled pressure—and life of a sort.[14]

Relating these two experiences involving "the bigger question of life itself" to Saunders, Fred worried that his letter was "morbid," but "I am trying to be honest."[15] What had bothered him after observing his grandfather was the realization that the objective distinction between life and death was *only* the difference between reflex action and no reflex action. His grandfather's death exemplified the lawfulness of the physical world. But Fulton's operation had revealed the vicissitudes of random accident; a slip of the physician's hand could determine life or death. The world seemed both determined and accidental; and an awareness of this incongruity would remain essential to Skinner's view of the world and its organisms.

There was, he told Saunders, a hierarchy of honesty in the way people recognized reality: "I could feel better about life if I'd close my eyes. But if I completely close them—Edward Guest—while if I squint just a little—a college professor of literature—[John] Galsworthy—H. G. Wells—or if I wear colored glasses—Catholicism—G. K. Chesterton—Thomas Aquinas—or wide open—nothing." To see the world with open eyes was to see a world of physical reflex and accident, a world not fashioned by the mind: "We go on thinking—yet do we live by thinking?—not by a damn sight. . . . Do you think Socrates drank hemlock so gracefully because he thought it out and found that to be the *right* thing? He did not—he *liked* the beau geste."[16] Fred had discovered a major paradox of behaviorism before reading the behaviorists. He had discovered that "by using his mind, man reduces mind to behavior."[17]

SKINNER PROBABLY FIRST READ about John B. Watson, the founder of American behaviorism, in the August 1926 issue of *The Dial,* one of the "little magazines" he read that published original fiction and poetry, critical articles, and reviews. Among the latter was Bertrand Russell's review of C. K. Ogden and I. A. Richard's *The Meaning of Meaning,* in which Russell made favorable reference to Watson. In a later issue Russell again praised Watson while reviewing E. A. Burtt's *The Metaphysical Foundations of Modern Science* (1925). More than any other writer, Russell was responsible for introducing Skinner to behaviorism as a philosophy of science.[18]

The editors of *The Dial* were sensitive to the role of science in the post–World War I era, especially to philosophical and literary attitudes toward science. Some writers during the 1920s attacked science as the agency of dehumanization and tended to be pessimistic about the future of a humanity dominated by science. Their disillusionment with science was influenced by the mechanized slaughter of millions during the war.

One well-received book criticizing science, Joseph Wood Krutch's *The Modern Temper* (1929), claimed that science was responsible for the sharp cleft between feeling and thought in the modern mind.[19] Krutch would later write one of the first widely read critiques of Skinner's fictional attempt to use behavioral science to shape a culture in *Walden Two.*[20]

On the other hand, a few writers, most notably Bertrand Russell and the literary critic I. A. Richards, argued that if science was used with intelligence it could benefit humanity, both materially and artistically.[21] Although Skinner would share such fashionable traits of young American intellectuals during the 1920s as cynical criticism of conventional middle-class values and reverence for artistic freedom, he was never anti-science. His modernist revolt against the bourgeois culture of his parents was never complete. Indeed, much of bourgeois culture was pro-science.[22]

Skinner did not remember reading Watson firsthand until the spring of 1928, after abandoning writing as a career. He emphasized that he had come to Watson through Russell, and he did not read Russell's sustained analysis of Watson's *Behaviorism* (1925) until early in 1928. Later Skinner wrote in a short essay, "Books That Have Influenced Me," that even though he had read Watson's *Behaviorism* (but exactly when he did not say), he was not sure whether he had ever read Watson's *Psychology from the Standpoint of a Behaviorist* (1919).[23] Indeed, it was the writing of his critical review of Lewis Berman's *The Religion Called Behaviorism* (1927) for the *Saturday Review of Literature* in early 1928 that Fred recalled as the time when he first defined himself as a behaviorist—admittedly without knowing very much about the subject: "[I] attacked the book as if I knew what I was talking about."[24] But his letters to Saunders show that his shift to the behaviorist standpoint occurred very early in his Dark Year—in the summer and autumn of 1926. The experiences he had undergone, as well as his reading and failed efforts at writing, all contributed to his adoption of behaviorism as a philosophy of science.

Fred's keen power of observation was already like a scientist's, though he did not yet *recognize* himself as one. He isolated phenomena, ignored irrelevant conditions or mental states, and focused on physical acts of the human organism. The crisis he called "dying from a congestion of ideas"—his questions about objective writing, his difficulty in finding a philosophy of life, the question of life and death itself as determined by natural laws and accident—was an intellectual malady eased by embracing a philosophy of science that recognized reality as determined reflex action. Scientific objectivity as expressed from the behaviorist standpoint became an alternative to the sentimental and traditional interpretations of mind, life, and death—traditions that were more comforting but less honest than behaviorism.[25]

Skinner's objective/behaviorist perspective did not appear suddenly in 1926 as if by magic or religious conversion. He worked toward it by recognizing that he could not be the kind of writer he admired, by closely observing the physical actions of living things, by reading Russell on Watson, and by using Saunders as a sounding board—by "making do." He judged literature and philosophy; he incorporated and discarded; he observed and recorded; but he never *followed* an intellectual approach or school. He found some important allies during the Dark Year, but they were never more essential than his own efforts to reach an objective interpretation of reality.

OTHER CIRCUMSTANCES made the Dark Year unpleasant, if not miserable. His old problem, "the conceit of the insecure," resurfaced.[26] He felt implied pressure from others to secure a paying job, which made it difficult for him to maintain self-esteem.[27] He was also socially uncomfortable, particularly at the dances given by the families of eligible girls: "I was outside the groups who arrived, their dance-books already filled, from pre-dance dinner parties."[28] The social isolation in Scranton exacted a heavy toll: "I have a feeling that I shall not survive unbroken."[29] In addition, his parents never let go of their original skepticism, so they were not supportive of the leisurely life-style of the would-be writer who lived under their roof.

Out of the house, Fred occasionally accompanied Dr. Fulton and his Scottish terrier, Pep, on treks around Lake Scranton. Though Fred considered Fulton a somewhat ridiculous character, he recorded the doctor's rationalizations of Pep's behavior. They, too, reveal Skinner's early behaviorist perspective:

> The doctor and his dog are becoming something more than idle amusement. . . . Actually the doctor has a sort of monomania. Back of it lie seventy years of practical, emotionless living, disappointment in his family life, the death of his only boy child, and the approach of mild senility. He is obsessed now with this dog, . . . [who] is unattractive, timid, mentally stunted. . . . He becomes a

> fool in the eyes of everyone but me when he attempts to justify the dog, or interpolate thought into the dog's action. He [insults his business associates by] breaking into their conversation with some pointless anecdote of Pep's colorless behavior. He takes the dog into the hospital with him and while there pays more attention to it than to his patients.[30]

Fred wrote to Saunders of how the doctor would ascribe mental processes to the dog:

> [Pep] comes up to us wagging his tail—"He says 'throw me a stick,' " says the doctor. . . . Lately I've got the habit too. It's quite fun to make up mental processes to fit a dog's every move. The conflict between mine and the doctor's is sometimes interesting. If we come to a parting in the path the dog will wait for us.
>
> "He doesn't know which way to go," I say.
>
> "He's waiting for me, aren't you, Pep?" says the doctor with a touch of pride.[31]

Clearly, Skinner was quite aware of the distinction between inventing mental states and the actual behavior of the dog, while Fulton was imagining a mental reality for Pep.

Although these outings provided Fred with some lighthearted moments, he remained vulnerable to his parents' criticism and Scranton's disapproval. But he did not openly rebel or immediately plan his escape. "My family ties prevent me," he explained, "not because I have a great deal of devotion and respect for my father and mother, but because they have suffered very much in the last four years and because my leaving them would increase their present anxiety to an unbearable degree."[32] Although he attempted "the pose of music critic," dashing off "musical criticism now and then for the *Scranton Sun* at four dollars the column," both pose and occupation proved unsatisfying.[33]

In the autumn of 1926 his father made a proposal that superficially seemed a solution to both his own and his parents' unhappiness. He offered Fred a very well paying job as manager of an employers' insurance bureau. But no such agency existed—it was William's ruse to guide his son off the writer's path, a detail Fred failed to mention to Saunders. The effect was to bring Fred to a career crossroads. Certainly he could no longer justify his plans to write, yet he hesitated to take his father's offer and embrace the materialistic and conventional life-style.

But no immediate career decision was made, and in early November

1926 Fred visited Percy Saunders in Clinton. Saunders recommended that Fred read H. G. Wells's new novel, *The World of William Clissold* (1926), which criticizes conventional social arrangements and urges that society be rearranged along scientific lines—a general idea, though differently stated, Skinner later applied in *Walden Two*.[34] He admitted being saved from accepting his father's offer and its allure of a materialistic life by reading another book Saunders recommended to him, Sinclair Lewis's *Arrowsmith* (1925).[35] This story describes an idealistic medical student who, after considerable struggle, devotes himself to pure scientific research. Like *Main Street* and *Babbitt, Arrowsmith* was one of Lewis's blistering critiques of conventional America, especially conventionality in academic and university life.

Fred especially admired the character Max Gottlieb, who successfully resists administrative demands for research with immediate practical results. Gottlieb reminded him of Hamilton's best: "a combination of Bugsy Morrill [Albro Morrill, his professor of biology], Bill Shep [William Shepard, his professor of French] and you, if you will allow me," he wrote Saunders. Like the novel's hero, Martin Arrowsmith, Skinner faced a choice between his ideals and settling into comfortable mediocrity:

> Hasn't Sinclair Lewis caught pretty well the inevitable struggle to choose between a reasonably smug conventional life and the chaotic road to being HONEST with your self? And hasn't he damn well represented the effect a comparatively trivial matter may have in deciding it? A girl, an old inbred desire to be thought well of, a love of "polite society," or any one of a dozen small desires takes on gigantic power to throw a decision the WRONG way.[36]

He admitted that "I think I was planning my immediate future so that I could soon marry," but that this had been a "pretty misty desire." But he also realized that his plans were vague, though desire was strong: "Not that I have any special person in mind; the desire to love is the main thing, . . . and a person can be on the verge of getting married without thinking about any prospective bride." Being without an object, this desire for matrimony most likely reflected Fred's loneliness and isolation in Scranton. He recognized that he, too, could "rationalize beautifully to justify myself just as the doctor rationalizes to explain some antic of his dog."[37]

Arrowsmith also suggested a curious parallel between fiction and life. Lewis had modeled Max Gottlieb on the German-American physiologist Jacques Loeb, whose attempts to create primitive living organisms from chemicals through parthenogenesis had created a sensation, and whose book, *The Organism as a Whole* (1916), Fred had read at Hamilton on the recommendation of his biology professor. Loeb, in turn, had been a student of Ernst Mach, the Austrian physicist/philosopher who had argued rigorously in *The Science of Mechanics* (1893) and *Perception and Error* (1905) that science should avoid all metaphysical assumptions.[38] Skinner maintained that his own intellectual genealogy could be traced from Mach to Loeb to the Harvard physiologist William Crozier.[39] Crozier had studied under Loeb and would later encourage Fred to experiment with tropism, a specialty pioneered by Loeb.[40] Skinner's later preference for studying the movement of an intact organism was fundamental to his scientific approach, as was his strict avoidance of metaphysical hypothesis in the scientific analysis of behavior.[41]

Arrowsmith did not make Fred a scientist, but it touched his life at a critical moment and, in championing a particular scientific ideal, represented a scientific, experimental tradition he would soon embrace. The book was also a vivid reminder of his ambivalence about the future, as it once again emphasized the old conflict he felt between his parents' dull conventionality and the alluring eccentricity of Percy Saunders. The tension he felt between the kind of life he wanted and the kind of life his parents wanted him to have had still not been resolved.

IN SEPTEMBER 1925 something happened that further complicated Fred's Dark Year. Harris Torrey, general counsel for the Hudson Coal Company, died unexpectedly and William Skinner assumed that he would be promoted to Torrey's position. But William had been in Scranton for only three years, and his legal experience for the company had been limited to cases involving workmen's compensation. To his dismay, he was not appointed general counsel.

While mulling over his future with friends, William then decided to enter private practice. He had hired his former secretary and opened a downtown office on New Year's Day, 1926. But at fifty, and not especially well known in the county—whose prosperity depended on a coal industry that was declining—William "spent hours simply waiting for someone to come in with a case."[42] He had saved enough money to avoid an immediate financial crisis, but he gradually fell into a depression.

The move to Scranton had been a terrible mistake for William. His favored son had suddenly died; he had failed to be promoted; and after nearly a year in private practice, he barely had enough business to warrant keeping the office open. His prospects slipped further when one of his few clients accused him of being dishonest about his fee and threatened disbarment proceedings. To make matters worse, Fred spent his time at home tinkering at writing and playing the piano, "with no sensible career plans at all."[43] He was especially concerned when he observed Fred sitting completely motionless in a chair for nearly half an hour at a stretch. This was another expression of Fred's detachment—a defense, perhaps, against emotional pain. He described his behavior as "a kind of existential state. . . . Somehow I become a separate world. . . . I seem to be free of time. The world seems merely an occasion—and one which is at the moment not acting on me."[44] He simply removed himself, mentally, from his surroundings. This was, in effect, a psychic Skinner box. At one point William suggested that Fred see a psychiatrist, but nothing came of it. It was not an emotionally or psychologically healthy time for anyone in the Skinner family, and it is no exaggeration to call 1926 the Dark Year for William as well as for his son.

Grace Skinner was afraid her husband would take his own life, but Fred believed his father's faith in progress would prevent him from committing such an act. One day William came home for lunch and went to his bedroom and wept. Believing his father's depression was due largely to his own idleness, Fred promised him he would get a job and start to earn money.[45] He knew his father would have loved to go back to Susquehanna, and probably would have if it did not mean admitting failure.[46]

Certainly his son's inactivity troubled William. Perhaps, too, he resented his days at home with Grace. Fred believed that his mother thought her husband was jealous of her practice of massaging Fred's scalp as he lay reading on the sofa in the library. Even trivial affection took on exaggerated importance, given Grace's sexual inactivity with William. Fred also believed his father resented his comradeship with Dr. Fulton. He took walks with him and washed and polished the doctor's Packard, something he never did with his father's car. On the contrary, Fred once carelessly drove off in the family sedan with one of the back doors swinging open freely. The door hit a tree and was nearly ripped off.[47] William's reaction to this negligence was more despair than anger.

Skinner remembered only his father's unhappiness during the Dark

Year, never any aggressive confrontation about Fred's idleness or threats to renege on the promise of a free year to write. Yet, as his letters to Saunders clearly showed, he felt and resented his father's displeasure in him and remained deeply ambivalent; he wanted both to please and to ridicule his father—but he always ridiculed covertly. Later, when Skinner was considering writing a novel about his early life, he referred to himself in Scranton as "the messiah, the fake and the solid achiever."[48] He wanted to do great things; he wanted to please his parents; but, he was, in fact, doing very little and considered himself something of a phony.

Soon, however, the fortunes of both father and son began to improve. In the fall of 1926 William invited to dinner the president of a small, local coal company, and he became infatuated with the Skinners' maid. In the following months there were secret liaisons and phone calls when he knew William and Grace to be away. Fred, however, was aware of the affair from the beginning and wrote Saunders cynically about "the Tired Business Man" and his romance.[49] Though the affair led to nothing permanent, the suitor began to send considerable legal business William's way. By early 1927 his decision to go into private practice looked much better. The following year he was elected president of the Scranton Kiwanis Club, a sure sign that he at last had arrived.

Meanwhile, Fred had begun writing a story about a woman who becomes religiously fanatic. He called it "For a Place in the Sky" and managed about "fifty words an hour on it"—a pace he found discouraging. More important, he had found a new activity that engaged his manual dexterity and provided him with a potentially profitable hobby. He had fashioned a work area in the garage where he could build model ships for possible sale. "I'm hoping," he explained, "that may be a way out for me. I've always been more or less adept with my hands, not exquisitely enough to be an artist, but perhaps clever enough to be a good carpenter."[50] He was returning to the hands-on inventive work he had found so rewarding as a child in Susquehanna—perhaps remembering his Grandfather Burrhus's skilled carving and finding he shared the knack.

Skinner maintained that the shipbuilding was an "escape" from writing.[51] Yet it was an escape he thoroughly enjoyed. And though no one bought his first ship, it "was worth a thousand dollars' worth of thrills for me to make. I get the same thrill out of making this boat that Doug Fairbanks must have got out of making his picture, *The Black Pirate.*" Long after his writing had succeeded and his books had brought him fame, Skinner still tinkered and improvised, and once even sculpted a ceramic head of his daughter Julie.

BY THE WINTER OF 1927 Fred had established a working schedule, a pattern he would continue all his life: "Every morning I read and study and write. . . . there's one interruption only: my mail is brought to me. Except for that I see no one; just write and think and judge what other people are thinking about." In the afternoons "I work at the bench . . . making [the ships] is great fun." He planned to recreate the phantom ship from Coleridge's *Ancient Mariner.* When completed, it would "give the effect of a half eaten carcass, the ribs of the boat exposed in parts, the sails in threads and the ropes thin silky webs. Colored in greens and blues and bronzes, all grayed as if in a mist."[52] And he had plans for other model ships. His letters to Saunders mention no stories being planned or written after early January 1927.

If writing failed to give him what he wanted, reflecting on his reading increased his affinity for science. His reaction to an article by the Italian philosopher-historian Benedetto Croce in *The Dial* in a letter to Saunders showed his enthusiasm for an antimetaphysical scientific standpoint: "Croce knows well enough . . . that science has displaced philosophy . . . , but, true to his tradition, he finds that the PHILOSOPHER of the future MUST BE a scientist. As if he reviewed evolution and said 'the ape is dead, and must now be a man.' " Fred agreed with Croce that "philosophy must drop the metaphysical and the closed system and turn itself to experience." That, however, was a clever way of trying to preserve philosophy. "But then you can't quarrel with [him] for wanting to perpetuate his species."[53] As his hopes for becoming a writer diminished, Fred's interest in the philosophy of science grew. He became interested in writers, like Croce, who abandoned metaphysics for scientific realism. Yet he saw a difference between being a scientist and being a philosopher who approved of science; one who still identified oneself as a philosopher could never really understand the world *as* a scientist. Perhaps that is one reason why Skinner never closely identified with American pragmatism, since that persuasion, however critical of traditional metaphysics, retained the philosopher's standpoint.[54]

Enthusiasm for the objectivity of the scientist encouraged his already detached, cynical view of himself: "As B. Frederic Skinner, slightly cracked, sits back in his artistic study, calmly reviewing the world as it passes by his window (slightly dusty) it has occurred to him that each is right in his own light (slightly platitudinous) and that to be militant against wrong is to spear windmills (slightly trite)."[55] He posed as a spectator who understood the relativity of values, but he would have quickly abandoned intellectual neutrality for the right cause—one in which he felt he could lead or at least be original. His detachment disguised his desire to be immersed in an activity that could totally absorb him. Indeed, the "B. Frederic" itself began an identity transformation. By the early 1930s he would become simply "B. F. Skinner"—an identity with an objective science finally found.[56]

He envied writers who had succeeded where he had failed. He read *Ulysses* in awe and claimed he would "feel myself a complete success if I could have written that book." He developed a special kinship with James Joyce, Ezra Pound, and Ford Madox Ford. Although there are great differences in style and subject among these three writers, it is interesting that each insisted on accurate observation—whether the stream-of-consciousness writing of Joyce, the imagist "thing-in-itself" poetry of Pound, or the social detail of Ford's novels. All championed what Fred called "pure" or "objective" literature. But his growing awareness that he himself could not produce literature he could admire led him to admit that it was time to "break from the family and set up a living in which I can respect myself."[57]

The humdrum routine of Scranton life was enlivened by a weekend trip to New York City to visit a former Hamilton classmate, Alf Evers. Evers had been one of Edward Root's most talented art students, and was now drawing at the American Art Students' League in Greenwich Village. They enjoyed the immense cultural offerings of Manhattan, and the visit made living in Scranton even more intolerable. Skinner later credited Evers with helping him to decide on a scientific career by maintaining that "science is the art of the twentieth century."[58]

In early April, a local gardener offered Fred a job as a day laborer, and he spent the next two months out-of-doors mowing lawns and planting shrubs. The work toned his body and lifted his spirits. In tribute to his liberation he composed a poem for Saunders, "Hymn to Labor: Action as the Solution to Doubt," which showed his determination to break the doldrums of his Dark Year:

(This, the farewell to my discontent!)
Awake, my soul, from dull seizure
Of the sweet hasheesh of too much leisure
And of the sin of sorrow now repent!
(This, the farewell to my love of sorrow;
This, the God-be-with-you to my doubt,
Which henceforth I must learn to do without!)
Now rings the wild Alarm! It is Tomorrow!

He wrote Saunders that he was enjoying *The Peasants,* by Wladyslaw S. Reymont, *Erewhon,* by Samuel Butler, and *Elmer Gantry,* by Sinclair Lewis.[59] These books had in common a rural ambience and characters who were active rather than intellectual. Interestingly, physical labor would be required of everyone in *Walden Two,* and Skinner himself later did gardening and isometric exercises to stay in shape.

Despite the invigorating outdoor work, Fred had resolved nothing about his future; indeed, a new sense of alarm appeared: "It seems that the rate of change in me is accelerating in geometric progression. . . . Is there any limit to the speed of mental metathesis?" Perhaps more than at any other time in his life, he felt that he had lost control of the direction in which his life was headed, that he was "being whirled somewhere and I'm not even enjoying the trip."[60]* A similar sense characterized many American intellectuals in the 1920s. They saw themselves as victims of forces that had in fact freed them to indulge in life-styles of their choice.[62] On a personal level, self-control would be important to Skinner. Periodically, throughout his life, he would "take stock," enumerate his goals, and evaluate his progress. Later, when he had a science to explicate and promote, this may have been a strategy he employed to prevent directionless, whimsical pursuits. But in 1927 he was plagued by doubts about his writing career and his future. He would maintain when he was older that chance had determined many of the major outcomes of his life—the college he attended, the woman he married, even his scientific discoveries. But he drew up career goals as a graduate student and planned his future with dozens of "Stock Taking" notes.[63]

*The historian Donald Meyer has commented on the sensation of being whirled toward an unknown fate. Meyer suggested that "whirling' . . . was the condition of illumination. But Skinner's sense of a loss of control over his future did not attach itself to any definite creative project, and so was perhaps more emotionally disturbing.[61]

During the spring of 1927, he sent Saunders his literary obituary. Again written in the third person, this tale of literary death was told by a detached observer. The subject was a young man who had "an Idea that no one seems to have had before." But rather than concentrating on his task, the young man became absorbed in the place his book would have in history, in literary reviews. Unable to communicate his Idea, "the young man took heart and killed himself. And he was not greatly disturbed that his Idea would die with him." "I was never anywhere near writing a novel," Fred admitted. "I don't know why I went on and on.[64]

A FEW MONTHS LATER he was forced to quit his gardening job when he suffered an acute allergic reaction after being pricked thousands of times while carrying barberry bushes for replanting. He decided to take a short vacation in Bread Loaf, Vermont, which would also give him a chance to get away from Scranton and his parents. But first he would take up an offer of his father's he thought would give him a financial stake in an independent future.

William agreed to pay him several thousand dollars to work on an important project. The coal companies were anxious to have a ready reference work that encapsulated the many, many legal cases that had arisen since President Theodore Roosevelt had created a board to adjudicate conflict between them and their unions. For most of the period from June to December 1927, Fred worked steadily on reading, abstracting, classifying, and cross-referencing hundreds of legal decisions. *A Digest of Decisions of the Anthracite Board of Conciliation* was privately published early in 1928, but Fred received no money for it for nine months and continued to rely on advances from his father.

Around the time he finished the project, he made a decision to enter graduate school in psychology. He had been especially influenced by an article by H. G. Wells in the *New York Times Magazine* in which Wells chose to save the man of science, Ivan Pavlov, over the man of literature, George Bernard Shaw, if he had but one life preserver. Fred probably made his decision about two weeks before he read the article, but it acted as a strong reinforcement.[65]

Late in October he went to Clinton again to seek advice about graduate schools specializing in psychology. Saunders, whose brother taught physics at Harvard, recommended that university. So did his old biology professor, Albro Morrill (several of whose students had gone on to Harvard Medical School), and President Ferry. Saunders and Morrill were well pleased with Fred's new direction. Though he would not apply for admission to Harvard for six months, his parents felt "tremendous relief." William was so pleased that he offered to take the family to Europe the next summer. Fred accepted, but only on the condition that he be allowed to spend a month or two there on his own before joining them.[66] The decision freed him to make an immediate change, and by February 1928, he was living in New York's Greenwich Village.

Before he visited Evers there the previous January, Fred had attended a lecture on Greenwich Village, at Scranton's Century Club. The lecture was given by a woman whose acquaintance Fred made. She and her husband invited him to visit, which he did a few weeks later. Attending a party hosted by this Village couple, he met Clara, who lived nearby with a roommate.[67] It was into this apartment that he now moved for his first few weeks in New York.

Greenwich Village had long been the New World haven for free-lance artists and intellectuals, the American home of "the lost generation." Depending upon whom you asked, the Village was a charming residential district, an Italian ghetto, a tourist attraction, a homosexual haunt, a never-ending party, or an art colony.[68] It appealed to Skinner because he found himself among people who embraced unconventionality, people who, like himself, had bourgeois backgrounds but were self-consciously seeking an alternative culture.

Fred kept his unconventional living arrangement a secret to his parents. Clara was married to a serviceman stationed in New Jersey. At some point after he moved in, "she began to fancy me. . . . I am sure I was not a satisfactory sexual partner. In fact, when we broke up, she told me as much."[69] Years later, upon hearing Gershwin's "Rhapsody in Blue," music associated "with a new, Bohemian and romantic life," Fred would recall "my life at that time—hours alone in the apartment, my reading, the Cézanne print, the groceries and kitchen, Clara and Doris, . . . early Sunday mornings."[70]

For a time Fred shared lodgings with his Hamilton classmate John Hutchens. While the apartment was pleasant, the room heater brought out bedbugs. So he moved back to Barrow Street, where Clara lived, but this time to an apartment closer to the river, where he stayed until leaving for Europe in July. His funds depleted, he decided to look for work. After considering selling theater tickets and writing reviews for newspapers, he took a part-time job as a clerk in the Doubleday Doran Company bookstore in a department store on Fifth Avenue. The salary was low and the hours passed slowly, but he did get some good books out of it. One was Aldous Huxley's *Proper Studies,* in which Fred saw Huxley also turning away from literature and toward psychology.[71] And he did find time to read, among other things, John B. Watson's *Pychological Care of Infant and Child* (1928). But his description makes clear that the job was not for him: "I sell books," he wrote Saunders, "suitable for a girl of thirteen who has just had her adenoids and tonsils removed and for elderly ladies."[72]

Life in the Village magnified certain cultural trends of the late 1920s. Fred recalled discussions at parties of Freud and psychoanalysis, a great pastime among artists, writers, and intellectuals.[73] Talking psychology took its place with other Village fads such as casual sex, discussing political radicalism, and vegetarianism. This was a novel culture, and he wanted to experience all the forbidden practices—he even hypnotized Clara once!

But eventually he tired of this life. Writing to Saunders to say he had been accepted at Harvard and was about to leave for Europe, he explained:

> New York has been a rich experience; but I am sated with it. There has been much uneasy idleness, a good deal of hard work, a great deal of idealized love and much plain sex. But for all it has given me (music, art, experience, contacts, variety) I am goddam sick of it. I have read, studied, worried, sunk into ecstasy, wallowed in depression, languished in boredom. If nothing else comes of it, I am satisfied that I shall never feel I am missing life, if I grow to live [old] quietly.[74]

When Fred left New York for Naples aboard the *S. S. Columbo* on July 2, 1928, he was not, as were some young American intellectuals who counted themselves members of the lost generation, seeking exile in Europe. Rather, he traveled as a self-conscious American tourist, thinking of "what Sinclair Lewis, H. L. Mencken or Ezra Pound would say about me." Skinner recalled that he tried to act as though he had made the journey many times and "must have seemed thoroughly obnoxious."[75] Insecurity always brought out the worst in Fred Skinner.

He visited the usual Italian attractions and then made short stays in Vienna, Munich, Basel, and Brussels. Fred had no qualms about using the fledgling air service. It had been only a year since Lindbergh had completed his solo flight across the Atlantic. On a flight from Switzerland to Belgium, he asked to share the open cockpit with the pilot and arrived drenched by a driving rain. Air travel was a way to escape traveling with the run-of-the-mill crowd. But it was also a chance to experience the dramatic technology of physical liberation he had been enamored of when he graduated from high school. To fly may have been an exhilarating reminder of his new freedom from old restraints.[76]

He joined his parents in Paris at a small hotel on the Left Bank and again, in their company, became an average tourist, taking in the Arc de Triomphe, the Eiffel Tower, the Louvre, and the Folies Bergères—the nudity here was tolerable to his mother since the show was a traditional part of the Paris tour. But Fred did not tell his parents about the other Paris he visited with his old Hamilton classmate, Jack Chase, who was living in Paris and translating French novels. After an evening of heavy drinking, Fred ended up with a prostitute.*

Fred convinced his reluctant parents to fly over the Channel to London. William hired a chauffeur to drive them to Devonshire, where he unsuccessfully tried to locate traces of the Skinner family. By late August they were aboard the *S. S. Harding* en route to New York, lodging in the first-class cabins insisted upon by his mother, ever concerned about what even complete strangers would think.

To pass the time, Fred recalled reading from several of the Henri Bergson books he had purchased in Paris. He was reading one when

> suddenly I was startled by a very large blast of a bugle. A member of the crew had come up behind me and had taken this customary way of announcing that dinner was served. After dinner I came back and began to read again. I went down the same page, and as I approached the point at which I had heard the blare of the bugle, I could feel perceptual and emotional responses slowly build up. The very thing Pavlov would have predicted.[78]

It was this conditioned response, not Bergsonian psychology, that really interested him.

Far more than the experience of Greenwich Village or Europe, the Dark Year in Scranton turned Fred toward behaviorism. Indeed, unlike

*The evening was not chivalrous. Chase had provided Skinner with a date, whom he left to pursue the prostitute; he ended up in a seedy hotel, unable to perform sexually. He had barely enough money to pay the cab fare to his parents' hotel.[77]

many American psychologists, he was relatively untouched by his travels in Europe. His predisposition for behaviorism was soon to become a militant conviction. The lingering tension between pursuing his intellectual ideals and satisfying parental expectations of conventionality would be resolved in an unexpected way. He was about to discover that he did indeed have something important to say.

4

The Birth of a New Science

The definition of the subject-matter of any science . . . is determined largely by the interest of the scientist. . . . We are interested . . . in what the organism does.

—B. F. Skinner, "The Concept of the Reflex in the Description of Behavior" (1931)

When you run into something interesting, drop everything else.

—B. F. Skinner, "A Case History in Scientific Method" (1956)

In September 1928 a slim, blue-eyed, twenty-four-year-old Fred Skinner walked from his rented room at 366 Harvard Street across Harvard Yard to Emerson Hall, where his courses were taught and the psychology laboratory was located. A fellow graduate student described him as "independent . . . self-starting and inventive" as well as "curious, ingenious, alive to detail."[1] Skinner quickly discerned a crucial difference between being a Hamilton undergraduate and a Harvard graduate student: "There is an air of informality about graduate work which was lacking in the undergraduate days. Here you either do it and get credit or don't and don't."[2] After finding living quarters, meeting professors, and beginning classes, he wrote Percy Saunders: "I am taking it easy my first semester. . . . After January 1 I expect to settle down and solve the riddle of the Universe. Harvard is fine. A strange and fearful freedom after Ham. Col. or Scranton. . . . If you ever come to Harvard be sure to call me. I should have, within a short time, some headless cats and a few conditioned reflexes on exhibition."[3] He had already set out a course of research in his notebook:

> It should be possible, by observing the development of an isolated specimen, to discover what part of the behavior of the individual does not depend upon imitations. It should be possible to further eliminate behavior resulting from chance reactions to environment by contrasting observations upon many specimens in differing environments. There must remain, it seems, a mass of behavior, a flow and reflow of activities, marked by definite purposes which is well enough described as inherited habit. The mechanisms of eating, copulating, fighting, nursing, etc. commonly called instinctive are abundantly untaught. Their survival value is so strong that we accept them as inherited equipment.[4]

Skinner enrolled in Physiology 5 with Hudson Hoagland; Physiology 20, laboratory with William Crozier; Psychology 11 with Carroll Pratt; Psychology 27 with Walter Hunter; and elementary German, to satisfy a language requirement.[5] His initial research, assigned in Hoagland's course, was a project measuring frog reflexes. Skinner proudly announced the scientific importance of his experimental work in a letter to his parents: "Right at present I'm putting all my time to getting ready for my great experiment. It will mean the construction of a great deal of apparatus, very accurate timing devices and recording machines." He would observe, within thousandths of a second, how long it would take a frog to jump after receiving an electric shock: "Every twitch the frog makes (and all the time he sitting on a plate without connections at all in order to make conditions as nearly natural as possible) will be recorded." But the observation of one of the Ph.D.'s that was the basis of his work proved unscientific. It turned out to be a case not of conditioned reflex but of lowered threshold.[6] He took great joy in finding something through experimentation that others had not expected him to find.

Professor Crozier's physiology course also involved laboratory work in measuring rates of reaction in intact organisms. In the late 1920s and early 1930s Crozier's laboratory had become a mecca for research fellows in the biological sciences. The place had an air of informality, adventure, and discovery. Moreover, "much of Crozier's work lay . . . between the frontiers of biology and psychology, and many of his students were drawn from the latter field."[7] He had a reputation for "kicking shins" and cultivated a no-nonsense, aggressive professorial style. He insisted on having students who had the confidence, talent, and determination to follow their own research interests. His best students were Gregory Pincus, who would invent the birth-control pill, and Fred Skinner.[8]

Like his mentor, Jacques Loeb, Crozier specialized in studying the movement or tropism of intact lower organisms, but he also worked with rats. Fred "carried on a small but careful" experiment in the lab, after which Crozier offered to help him personally to pursue it further.[9] Skinner was greatly impressed with Crozier during his first year at Harvard.

He also quickly sensed that his experimental ability could determine his professional future. Though he was looking for specific physiological reflexes rather than simply observing behavior, he was already focused on the intact organism rather than on a surgically isolated muscle or gland, as were Pavlovian behaviorists. Explaining his career goals in a letter to his father, he envisioned his future in a university setting. He mentioned nothing about reviving plans to be a writer, which no doubt pleased William.

SKINNER DECIDED TO FOCUS ON the field of physiology and psychology at a moment when historical developments had paved the way for a new behavioral science. Experimental work with the intact animal versus Pavlovian surgical preparation had gathered momentum in the late nineteenth century, paralleling advances in physiological reflexology. Darwin's *The Origin of Species* (1859) and *The Descent of Man* (1871) had called into question attributing mental processes to animal intelligence. Darwin, more than anyone else, had undermined the traditional distinction between human and animal intelligence as well as the related distinction between the supernatural and natural. Lesser figures such as Darwin's young friend George Romanes, and Romanes's friendly critic Lloyd Morgan, carried on Darwin's interest in animal behavior as natural science. Both Romanes and Morgan made a case for the mental capabilities of animals, but Morgan moved away from anecdotal reporting toward experimental techniques by using trial and error to ascertain the quality of animal intelligence. In 1898 Edward Lee Thorndike, an American psychologist, had introduced "puzzle boxes" to test animal intelligence and suggested that animals might not have ideas at all, only acquired connections between stimulus and response.[10] This was approaching conditioning as the study of the behavior of animals, although Thorndike did not deny that consciousness existed.

Meanwhile another scientific emphasis emerged: the experimental study of human consciousness as a science. In the 1870s laboratories for measuring the elapsed time between sensation and perception were established in Germany and the United States. Wilhelm Wundt and his American student James McKeen Cattell pioneered a "reaction time" or "brass instrument" psychology; others, such as William James and Edward Bradford Titchener, used "introspection" to study consciousness, respectively, as functional relations or the structure of mental elements.[11] Regardless of emphasis, introspection was an attempt to look inside the mind scientifically. Its proponents held that ideas emerged from or paralleled physiological processes in the brain; higher mental attributes such as perception and cognition were associated with physiological functions. By 1910, however, introspective psychology was on the defensive in America, challenged by younger men like Thorndike, Robert M. Yerkes, and John B. Watson, who were all interested in animal psychology.

The early-twentieth-century shift from psychology as the study of human consciousness to psychology as the study of animal intelligence signaled a more general shift to psychology as the study of behavior. In 1914 Watson announced that "psychology, as the behaviorist views it, is a purely objective, experimental branch of natural science which needs introspection as little as do the sciences of chemistry and physics." Dramatically proclaiming the death of psychology as the study of the mind, Watson advocated an alternative field, the study of behavior.[12] Behavior could be studied without reference to conscious processes, which were clearly nonobservable in both human and nonhuman subjects.

Skinner later believed that Watson had brought the "promise of a behavioral science," but this was not the same as delivering the science itself.[13] A bona fide science of behavior was first established by the Russian physiologist Ivan Pavlov, whose work became available in English translation in 1927. Pavlov worked with surgically prepared digestive glands to study conditioning. He spent thirty years measuring the secretion of saliva and, through his experimental findings, discarded the study of psychology as mental activity. While admiring Pavlov's skill in controlling experimental variables, Skinner would soon insist that the scientific study of behavior need not depend on physiology. In an early Harvard note he wrote:

> The argument against physiology is simply that we shall get more done in the field of behavior if we confine ourselves to behavior. When we rid ourselves of the delusion that we are getting down to fundamentals, when we get into physiology, then the young man who discovered some fact of behavior will not immediately go after "physiological correlates" but will go on discovering other facts of behavior.[14]

Indeed, Skinner would become a "descriptive" or "radical" behaviorist precisely because he denied that behavior is determined by processes within the physiology of the organism.

His approach to experimental work was characterized by mechanical improvisation and inventiveness. The Harvard psychology workshop located in Emerson Hall was an exciting new place for making do:

> I had never before used anything more complex than a vise, a hand drill, a hand saw and a coping saw, but the shop had a circular saw, a drill press, a lathe, and even a small milling machine discarded by the Physics Department. . . . All sorts of supplies were available; shelves of brass and iron wood screws and machine screws and nuts in Salisbury cigarette tins . . . and rivets, cotter keys, and small brass and iron pins in tins that once held Cuticura or Resinol ointment. . . . There were boxes of piano wire with which you could wind springs on the lathe, and shelves of strap and plate brass steel.

Such an array of machinery and odds and ends was simply irresistible. It became the center of Skinner's activity.[15]

Puttering in the shop sometime during his first year, Skinner made a gadget that would play an important role in the research that resulted in his greatest experimental discovery: "a silent-release box, operated by compressed air and designed to eliminate disturbances when introducing a rat into an apparatus."[16] This box was directly related to other devices and experiments that evolved into what was variously called the problem box, the lever box, or the experimental space, but which was most widely known as the Skinner box. Later he seemed ambivalent about this name, as its value was not in the architecture but in its results—a predictable rate of response as a measure. Although he proudly acknowledged the entry of "Skinner box" into the dictionary, years later at a professional meeting he suggested substituting the term "experimental space." Someone jokingly asked if it would be okay to call it a "Skinner space."[17]

DURING FRED'S FIRST YEAR of graduate school he came close to changing his field from psychology to physiology. His indecision was prompted by his belief that physiology was more scientific than psychology and allowed the kind of experimentation in which he was most interested. He explained his thinking to his parents in January 1929, emphasizing that the physiology department had better facilities and that it would be a good chance to ally himself with the influential Crozier.[18] To Saunders he had previously written, with no sign of ambivalence: "I am working as hard as I have ever worked, but freely with time and subject matter at my own choosing. I have almost gone over to Physiology, which I find fascinating. But my fundamental interests lie in the field of psychology and I shall probably continue therein, even, if necessary, by making over the entire field to suit myself."[19]

By February he had passed the German and French exams required for the Ph.D. He was invited by a professor at Tufts College to give a talk on insight to a class in animal psychology.[20] For the spring term he enrolled in a philosophy course and three psychology courses. His favorite was Psychology 20C—animal research, taught by Walter Hunter. It was a seminar in animal behavior that met once a week to discuss individual work on some aspect of behavior.[21] He would be working with two squirrels, which he promised his parents he would keep at the lab, not in his room.[22] Even before those two squirrels arrived from the supplier, however, he had experimented in Harvard Yard by dangling a peanut on a string from a tree branch to see whether a squirrel would pull the string to bring the peanut within reach. It did, but Fred never believed this proved the animal had used "insight," or mental abilities, to solve the problem. It was trial-and-error conditioning, not an unobserved mental operation.[23]

Walter Samuel Hunter was a professor at Clark University who occasionally taught courses on animal behavior at Harvard in the late 1920s. Considered an important behaviorist, he had a reputation as a careful, clever experimenter who questioned the widely held assumption that animals learned primarily through imitation. Skinner remembered Hunter posing a question: "A dog chases a rabbit. The rabbit runs, the dog runs. Is the dog imitating the rabbit?"[24] Hunter furthered Thorndike's work with puzzle boxes, problem experiments in which cats "learned" to escape through trial and error. Thorndike had found that stimulus-and-response connections govern the behavior of cats, dogs, and even certain monkeys. It was unnecessary to posit a reasoning process to explain why animals modified their behavior.[25]

Hunter, working with raccoons, pigs, and rats, devised more sophisticated problem experiments, building a release box and multiple-choice chambers in which subjects learned, for instance, to exit a chamber in which a light bulb had been lit. He is probably best known for his delayed-response experiments, in which animals were detained before being allowed to choose the correct door. The results of his discrimination experiments showed that explanations of behavior were more complex than Thorndike had assumed. Though he did not deny mental processes, Hunter was far more interested in animal behavior than in animal reasoning. He was described by another Harvard graduate student at the time as "a breath of fresh air." Fred recalled Hunter saying: "Skinner, it just takes one little idea to be a success in American psychology."[26] In terms of "the lineage of ideas," Skinner considered himself "a grandchild of . . . Hunter."[27]

With another graduate student, Dwight Chapman, Skinner investigated the insight of young squirrels. They were interested in testing the theories of the German gestalt psychologist Wolfgang Köhler. Köhler had done extensive work with apes on one of the Canary Islands and had maintained that apes did not solve problems through trial and error but through perceptual restructuring. He had challenged Thorndike's conclusion that animal problem solving is governed by trial and error, with accidental success. Experiments with imprisoned animals in puzzle boxes and mazes offered no opportunity for the demonstration of higher mental processes.[28] Köhler claimed to be proving experimentally that animals did more than behave; they could think as well.

Skinner was not convinced. He made fun of Köhler's work with apes by writing a spoof depicting a Köhlerlike experiment gone awry:

> A "scientist" in a white coat is seen pointing to . . . a basket hanging from a high branch of a tree on a long rope, some boxes to be piled by the ape to reach a basket, and a banana. . . . The scientist picks up the banana, climbs a ladder against the tree, and reaches for the basket. He slips, grasps the basket, and finds himself swinging from the rope. He begs the ape to pile some boxes under him so he can get down, but the ape refuses until the scientist throws him the banana.[29]

Chapman and Skinner considered, with Hunter's approval, submitting a paper for publication describing their work with the squirrels. But Chapman lost interest and Skinner later surmised that even though they had rejected Köhler, they "were leaving too much to the supposed mental processes of the squirrels"—to the problem-solving approach.[30]

In June 1929 Skinner learned of his appointment as Thayer Fellow for the following academic year, which "clinched my loyalty to psychology" because it came with a good-sized office.[31] His decision to remain with psychology also owed much to the arguments of a psychology graduate student, Fred S. Keller, who convinced Skinner that he could make a *science* out of the study of behavior and still get his doctorate in psychology.[32] That, indeed, was crucial because Skinner had never desired to become a psychologist in the sense of someone who studied the mind scientifically; he did not believe the "mind" existed.

Although Keller and Skinner discovered they had dissimilar tastes in music, sports, and women, they became best friends. Keller remembered that Skinner tried to help him with his dissertation project, a maze experiment, by inventing more apparatus.[33] They ignored Prohibition and occasionally drank and dined together. One night at the Commander Hotel in Cambridge, they discussed how problems of psychology should be selected. Keller argued that psychological problems issued "from what people needed most." Skinner maintained that psychological concerns "came out of data." Keller considered himself a late bloomer, being slower to accept Skinner's operant apparatus and slower to abandon "the old psychology." Skinner recalled that he had not been "indoctrinated in the old stuff" and did not know much psychology until he taught it in 1936, whereas Keller had studied it and had a part-time teaching position for a period while completing his graduate work.

According to Keller: "We were sort of stylish in those days. . . . We were quite elegant." Skinner agreed.[34] Having a colleague who shared an interest in the study of behavior and who was also a friend was important to Skinner: "I had little social life during the first two years of graduate work, but it took a gentle friend like Fred Keller to put up with me."[35] Skinner described his lifelong friend as "the philosopher of behaviorism, whereas I am the practitioner of the science of which behaviorism is the philosophy."[36]

By the end of his first year at Harvard, Skinner had learned that "Harvard University takes little or no interest in the private lives of its graduate students."[37] At Hamilton he had lived with others in a dormitory or fraternity house; now he lived in apartments. No one cared whether he went to chapel or teased him for being a newcomer. Valuing privacy, he found the Harvard environs accommodating. But his independence was not the libertarian freedom he had enjoyed in Greenwich Village. He set up a strict work schedule: "I would arise at six, study till breakfast, go to classes, laboratories, and libraries with no more than fifteen minutes unscheduled during the day, study till exactly nine o'clock and go to bed."[38] But Keller recalled that Skinner finished his experiments in the lab early and would play Ping-Pong in the afternoons, while other graduate students were still at the lab; this caused some irritation.[39]

Harvard students were judged on their determination and originality. Skinner informed his father, "I have gone at making associations rather slowly, which was fortunate, for I find the more important men are those whom you do not at first meet. However I now have a certain standing among other graduate students and I find my opinions (which I am careful not to offer too often) are listened to and respected."[40] His newfound independence gave rise to resentment of parental intrusion. "I don't mind phone calls," he wrote his parents, "so long as you pay the charges. But supposing I had gone . . . a couple of hours earlier and hadn't come in till eleven or so to get your wire? . . . I'm not scolding but it seems to me we're too far apart to count on keeping in day-by-day touch with each other."[41]

In the fall of 1929 Skinner was still living in the rented room on Harvard Street. In the cellar he kept a new purchase. Perhaps trying to remind his parents of their overprotectiveness, he perpetrated a gentle hoax:

> Well, I have bought a vehicle. The finish is dark blue and all the nickel is chromium plate. . . . It is really quite a snappy model—wire wheels and everything. I am sure you will not mind my getting it as I am a careful driver. The upkeep is really very little and I will save a great deal on car-fare and shoeleather.
>
> It is really a good bike (Good gracious, did you think it was an automobile?)[42]

On his bike he commuted to classes and the new psychology laboratory now located in Boylston Hall. Fall term he took Psychology 33, Perception, with Edwin G. Boring. He judged it "the least interesting course I ever took . . . and I got the highest grade in the exam by preparing twenty mnemonic sentences which permitted me to cite all the principal workers in all the fields we covered."[43] This was an early example of his effective manipulation of his own verbal behavior. Disliking the course, he enjoyed finding a way to master a topic of virtually no importance to him.

Much more interesting were two research courses and Crozier's Physiology 3, Analysis of Conduct. Crozier "enjoyed working with his hands as well as with his mind."[44] He must have been pleased to have a student so nimble of hand and mind as Fred Skinner. Recalling Crozier as "an aggressive person [who] made enemies [and] was an empire-builder," Skinner stressed that "he never tried to get me into his racket. He backed me up, he gave me space, he gave me money, he put me up for fellowships and so on [but he] never once said 'Look, why don't you do something with temperature coefficient' or something like that."[45] With Crozier, Skinner could be original, go off the beaten scientific path, without fear of reprisal or professional jealousy.

Throughout that second year at Harvard Skinner was absorbed in research, and he abandoned squirrels for rats as the subjects of his experiments. Maze experiments with rats had developed at Clark University through the efforts of Linus Kline and his graduate student Willard Small, as well as at the University of Chicago by the young animal psychologist John B. Watson. By the end of the first decade of the twentieth century the maze was routinely used to test sense discrimination and motivation and, by the mid-1920s, to test learning ability. The use of multiple animals became standard practice, with twenty or more rats making repeated maze runs. American behaviorists such as Edward Chace Tolman and Karl Lashley were among the notable rat researchers when the maze reached its zenith about 1930.[46] But Skinner found that the maze did not provide him with means to control the variables of an experiment so that the rats' behavior could be accurately measured and hence predicted. There were difficulties in separating the ability to learn from levels of activity and emotional excitability when rats ran mazes. Therefore when the same experiment was repeated, there were often different results. Skinner's criticism notwithstanding, the maze had distinguished adherents and was used into the 1940s and 1950s.

Skinner became interested in finding a way to measure and record his rats' changing postures. The physiologist Rudolf Magnus had shown how surgically altering the nervous system affects movement in cats and rabbits. Skinner wondered whether the behavior of the new generation of young rats he had acquired was similar to that of adults with impaired spinal cords. Here was an opportunity to study animal reflexes in the whole organism, and he decided to repeat some of Magnus's experiments. Skinner knew that juvenile rats did not maintain a normal body temperature when moved to hot or cold environments. Therefore, he could modify their environments and measure temperature changes; he could study "temperature coefficients of the reflex process."[47] He devised a large box with a glass window in front, a heater, and a thermostat. Sleeves cut in the side of the box allowed him to reach the animal. When the box proved too cumbersome to enable him to handle the rats, however, he abandoned the experiment.

Next he made slings on which he placed the young rats to observe their leg muscle reflexes. It was, however, difficult to record these specific leg muscle movements in a sling, so he constructed another apparatus. He mounted a postcard-sized, cloth-covered piece of aluminum on two parallel wires—a "piano-wire platform" that served as an experimental space upon which to study reflexes.[48] When a rat's tail was pulled, it reacted by clinging to the platform and moving forward. A single reflexive response could be observed under controlled conditions. To transcribe the observed result, he used a recording device called a *kymograph.* It was a drum that was turned by a motor equipped with a set of gears that could be turned at various speeds. The drum was wrapped with paper that had been blackened with soot ("smoked paper"). When a point scratched the paper-covered drum during an experiment, it left a white mark as a record. Thus, the action of a reflex could be represented by the kymograph marking. The smoked paper was then shellacked and dried for preservation. Skinner improved the quality of the record by using sheets of clear gelatin instead of paper; the gelatin could be projected onto slides, and copies could be made on blueprint paper used by architects. Now when he pulled the rat's tail, the gelatin-covered kymograph recorded the subject's forward movement: "My table and kymograph seemed to report such leaps fairly accurately and I was overjoyed. Here was the kind of thing I was looking for: the reflex behavior of an intact organism."[49]

By the summer of 1929 Skinner was using a modus operandi that would characterize his approach to experimental work throughout his life: "When you run into something interesting, drop everything else. I don't say ignore everything else, just drop it. It is not going to be as productive."[50] He was still making do as he had during his boyhood—improvising and inventing, using available materials, and pursuing what interested him, what reinforced him. A distinctive experimental style emerged. He proceeded not by continuing to use the same experimental format regardless of results, but by inventing new apparatus; apparatus that he abandoned if it did not produce results that could be reproduced without significant variation. He assumed he would eventually find and record orderly behavior, and the best way to do so was to construct the appropriate experimental apparatus. He was not so much experimenting to achieve desirable results as *devising* apparatus that allowed orderly results to be observed. The distinction is important because the most fundamental manipulation was in the invention rather than the experiment.* The invention of clever yet simple devices would eventually give him control over the behavior of his experimental subjects. Invention to achieve experimental control was the functional value of his talent for making do.

LATE IN 1929, though he had yet done nothing scientifically remarkable, he "began to be unbearably excited. Everything I touched suggested new and promising things to do."[52] The realization that he could design apparatus that promised experimental success meant that he could shape his own scientific future. Keller recalled that Skinner was "apparatus-minded" and overindulged himself—making devices, remaking them, changing them, tinkering with them.[53] Yet Skinner found inventing reinforcing, and he was on the verge of finding a vital connection between invention and achieving experimental control.

He had not abandoned his silent-release box, and he began to study the reflexes of a single rat as it entered a modified box—

> about two feet square with double walls of Celotex [an insulating material]. Inside it against one wall I put a small tunnel-like structure at the top of a flight of stairs. . . . I planned to study how [the rat] moved forward down the steps

*If Skinner was more fundamentally the inventor than the experimenter—although these roles obviously are related and do overlap—then his search for measurable experimental results depended on his ability to fashion a space in which such results appeared. What came under his direct control was not the experiment itself but *the making of it.* Here the crucial question was whether the essence of science was observation or control. For Skinner, as for Crozier and Loeb, control made science scientific; but the way to achieve control was through an experiment that properly controlled variables, an activity closely related to invention. The Anglo-American scientific tradition from the work of Francis Bacon through today has focused on controlling nature rather than merely observing it. American inventiveness in general and Skinner's in particular translated into technology because technology was by definition the most reliable way to control nature.[51]

> and how it pulled back when I made a noise. These were I thought reflexes. I watched the rat through what I hoped was a one-way window and recorded its progress out of and back into the tunnel by moving an arm, which pushed a pencil back and forth on a moving strip of paper.

He controlled the rat's movement in and out of the tunnel by clicking an old telegraph receiver key. The first click sent the animal scurrying back into the tunnel, where it waited several minutes before reappearing. Each subsequent click, however, had less frightening effects on the animal. "I hoped," he recalled, "to plot a curve showing this process of adaptation."[54]

But though a record appeared on the kymograph paper, there was too much variability in the rat's movement to leave a quantifiable pattern. How was Skinner to translate the recorded values into a measurable result? Proper control of the rat's movement, using the clicking sound as a stimulus, would, he hoped, result in a predictable response. He had not yet managed to achieve this goal, but he had devised an experimental, boxlike environment that allowed him to observe and record—albeit imperfectly—the special reflex behavior of one rat.

Next he let a rat out of his silent-release box onto "a ten-foot strip of light spruce on tightly stretched wires"—a jumbo version of his piano-wire platform for the baby rats. Now, rather than pulling a rat's tail to change its posture so as to observe shifts in muscle reflexes, he induced the adult rat to travel the length of the platform by delivering food at the end of the run. At the halfway point he sounded a click. A kymograph attached to the wired strip recorded changes in movement after the click. "I planned to measure changes in behavior as a rat slowly got used to the click," he wrote. "Perhaps I could even get it to stop in response to a conditioned stimulus." The conditioned stimulus was the clicking sound that accompanied the delivery of food. Again, however, the problem was too much variability in recording changes after the click. This time he traced the problem not to the rat but to a slight bouncing of the apparatus.[55] He replaced the stretched wires with vertical glass plates, but the kymograph still recorded too much irrelevant movement to leave an appropriately quantifiable record of starts and stops.

Skinner valued the spruce-strip experiment, for "it led me to watch a rat and try to account for its behavior." This was significantly different than studying a specific reflex. He described the animal's movement in terms of "adaptation" and "prepotency." It was not yet the language of his behaviorism—there was, for instance, nothing about reinforcement or discriminative stimuli. Neither, however, did he lapse into mentalism and discuss "insight" or even "motivation" or "drive." The central problem remained: "How was I to convert those wiggly lines into significant values?"[56] And how was he to acquire results that could be measured and replicated under the same experimental control?

He solved the problem of quantifying his experimental record accidentally. Skinner arranged experimental space so cleverly, and so quickly noticed changes in behavior that might be experimentally controlled, that he made it more likely for the happy accident to occur. For example, after his rat had made a run on the spruce-wired board and was fed, Skinner had to carry it back to the silent-release box for another run. This not only seemed inefficient but occasionally disturbed the rat. So he constructed a "back alley" along which the rat could return to the "straightaway" after completing a run. Eating the food would serve as a "stimulus" to make another run. There was "no reason," he later observed, "why the rat could not deliver its own reinforcement."[57] The experiment would be self-perpetuating. The addition of the alley, however, had an unexpected result: The rat did not always immediately begin another run after eating. The delay interested Fred more than the run, so he stopped making the clicks and began to measure the delay in eating with a stopwatch. Immediately he saw that it was not necessary to have a long runway to study the pauses. Perhaps the pauses themselves would yield a quantifiable result. He had begun to study the *time* between the behavior of eating and the behavior of running. Time became the variable he would soon be able to control experimentally.

He built still more apparatus, this time a three-foot-long track mounted so it tilted like a seesaw. The run had become a rectangular box. As the rat ran down the short track it tilted, and the tilt moved a small wooden disk with drilled holes. Each tilt of the track turned the disk so that a food pellet dropped into a cup. And each tilt was recorded by a kymograph mark—the time or pause between runs being the space between the markings. As the rats sated themselves on the food, they made fewer and fewer runs, which in turn made the kymograph marks farther and farther apart. By connecting lines from mark to mark, Skinner could graph the time between runs. He seemed at last to have found a reliable quantitative measure. He wrote his parents: "The experiment is coming on splendidly, new aspects cropping up each day which seem to indicate even greater importance than we thought at first. Crozier is anxious for me to write up and publish a first statement but I am waiting until I have checked every possibility of some technical mistake."[58]

Then another happy accident—perhaps the most important one of his experimental career—helped Skinner's progress. The disk of wood from which he had made the food dispenser

> happened to have a central spindle, which fortunately I had not bothered to cut off. One day it occurred to me that if I wound a string around the spindle and allowed it to unwind as the magazine was emptied . . . I would get a different kind of record. Instead of mere report of the up-and-down movement of the runway as a series of pips in a polygraph, I would get a curve. . . . The difference between the old type of record and the new may not seem great, but it turned out the curve revealed things in rate of responding and changes in that rate which certainly otherwise would have been missed.[59]

The thread on the spindle unwound rather than wound giving upside-down curves, but this was easily corrected. Skinner now had an invention that would be indispensable to behavioral science.

The curves soon yielded a cumulative record reasonably free from variation. The "cumulative recorder" provided a simple curve measurement, that is, rate of response, that would be seen as the rat ate, an "ingestion curve." "You could see changes," Skinner recalled. "The first [curves were] just slowing down when . . . [the rat] got fed up. But the tangent of the curve told you exactly how hungry the rat [was] at a given moment."[60] Rate of response was a simple relationship between time and eating. The tilt-board apparatus gave Skinner his first ingestion curves—curves that recorded the pause between each feeding of a dropped piece of food.

Considering these ingestion curves, he realized that the rat need not be on a tilted track. Indeed, it was not necessary to have locomotion apparatus at all. Once again, he modified his apparatus:

> I built a food bin with an electric contact on the door and installed it in a doubled-walled box, with a similar bin for water alongside. I deprived a rat of food twenty-four hours and then let it get its daily ration from the bin. The pellets were hard [he had started to make his own mixed-grain pills] and took some time to eat, and a session lasted as long as two hours. With the behavior thus reduced to opening a door, I began to get more orderly results. The rat

> ate rapidly at first but then more and more slowly as time passed. The cumulative curves were smooth.[61]

Even this arrangement was not exactly what he wanted, so he bent a wire into a lever that a rat could press down. Each press delivered a pellet from a glass tube instead of the wooden food disk. Now he was recording lever presses, the rat's response to the delivery of a food pellet, in the double-walled box as an ingestion curve. They continued to produce remarkably smooth curves. The cumulative recorder, Celotex-walled box, and lever press constituted the first Skinner box. At last he had constructed an apparatus that recorded behavior reasonably free from variation and could be repeated over and over as *lawful behavior*! Lawful behavior is behavior that under the same experimental conditions would occur again and again; by the experimental control of a rat in a Skinner box, he could *predict* the future behavior of the rat.

Skinner had just turned twenty-six and wrote his parents of "the greatest birthday present I got": "what heretofore was supposed to be 'free behavior' on the part of the rat is now shown to be just as much subject to natural laws as, for example, the rate of his pulse. My results seem to be very conclusive, and barring some slip-up in technique, are really important." This was his breakthrough discovery and, not surprisingly, "Crozier [was] quite worked up about it."[62] Skinner had found a quantifiable measurement of the behavior of an organism: the lever-pressing behavior of a rat. Rats in the same state of food deprivation would press the lever at the same rate of response, which produced strikingly similar ingestion curves on the cumulative recorder.

There was, however, yet another vital experimental surprise. Skinner now expected to get smooth curves showing rate of response, but "I didn't decide in advance that I wanted to prove that one reinforcement conditioned behavior. To my amazement I discovered it":[63]

> What I did was to control the conditions in the box following Pavlov's discovery that you have to control the conditions to get precision. One of the things I did [was] I put the rat in the box . . . many days and gave it just food to eat, so it got used to that. Then I operated the food dispenser . . . before the lever could be pushed. . . . The idea was I didn't want to disturb the rat when it heard the magazine. So finally you have got a rat that goes into the box, it eventually . . . picks up pellets whenever the magazine operates. What I didn't realize was . . . I was conditioning the reinforcer. The sound of the magazine. Two of my

> rats [pushed the lever] instantly. . . . Now that was a surprise. It wasn't what I was looking for. . . . I was disappointed because I wanted to get a learning curve.[64]

What Skinner had discovered was that the instantaneous sound of the magazine combined with the immediate appearance of food resulted in reinforcing conditioned behavior. This was neither a learning curve nor a learning process, only the effect of reinforcement on the rate of response. "If you just give an animal food, that isn't instantaneous. When you push down [the lever] and it goes BANG, that BANG is the thing. It is absolutely instantaneous with the movement, and that is what makes it possible."[65] With his tinkering, Skinner had accidentally made the crucial discovery of immediate reinforcement.

After recognizing the significance of instantaneous conditioning, he did not then hypothesize that he could introduce different schedules of reinforcement: "I didn't decide, 'well, you ought to get different effects from schedules of reinforcement.' I was running out of pellets, so I decided I would only reinforce now and then. Then I immediately began to schedule performance, and as it turned out many different ones."[66] No hypothesis was required; he simply found out that the rat would continue to push the lever at a different rate if fed on different schedules—prompted by a scarcity of pellets.

The simplicity of his apparatus made the experimental results all the more striking and convincing. Any carpenter could build a Skinner box and get the same amazing regularity and predictability the inventor had gotten. Skinner was proud of his invention of a new kind of behavioral technology. He had made all the equipment himself. And when he realized he could electrify his apparatus, including the delivery of pellets, he fashioned that equipment as well:

> I bought coils from Cenco Co. at $1. each. With them I made ratchets to wind string to lift a sliding scriber along the drum of a kymograph for a cumulative recorder. With them also I made my own relays, using galvanized iron, cut with sheers for frames and armatures, soldering on bits of silver (from ten-cent pieces) for contacts. I made several mechanical lock-up relays; one relay closed on a brief pulse and was held closed by a catch which could be released by operating another coil. So far as I knew, it was an original idea.

He admitted that "I am not unaware that the equipment is regarded as 'Skinnerian,' and [am] not displeased, I suppose."[67] The invention of the equipment was the essential technology that made his science possible.

WHAT HAD EMERGED WAS STUNNING, an apparatus that allowed the experimenter to predict far better experimental results than those with the maze. Skinner's science of behavior was predicated on the ability to control the results of his experiments.

As with his equipment, there was a remarkable simplicity in his experimental results. The rate of response curves that the cumulative recorder yielded could be read by anyone, and their regularity was astonishing—as if he had discovered a law of nature. There was no need for sophisticated quantitative manipulation or even a breaking down of data into smaller parts. There was no need to study, as Pavlov had done, an isolated and specific reflex, muscle twitch, or gland to achieve experimental control. The necessity of interpreting experimental results at all seemed to have dissolved. Skinner's behavioral science spoke for itself, because the cumulative records of different schedules of reinforcement did just that. As Keller recalled, "the excitement of Skinner was in watching what he said to be true turn out to be true."[68]

Skinner's experimental success in early 1930 coupled with his provocative operational definition of the reflex provided the initial opportunity for a new science of behavioral analysis. Although he would not complete the theoretical and experimental foundation of "operant science" until the publication of *The Behavior of Organisms* in 1938, the watershed year was 1930. Between 1928 and 1930 he had experimented in search of quantifiable order; after 1930 he found that much seemingly spontaneous behavior was determined by its reinforcing consequences and could be described as laws of behavior.

As his excitement and confidence soared, so did his weight. He told his parents that he was over 140 pounds—"as high as I have ever been."[69] He also knew he "could rest on my laurels" and coast toward his degree on the strength of the work he had already done. "But there is more to it and I am going to keep plugging."[70] Crozier's recognition meant a lot to him, but Crozier wanted no credit for Skinner's accomplishments. He once crossed out an acknowledgment Skinner had made to him on a paper he wrote, explaining: "We do not exact tribute here." Skinner learned the lesson and passed it along to his students: "You do something on your own, you ought to get credit for it."[71]

By the end of the academic year 1929–30, Skinner was making a name for himself in the new "camp" of behaviorism. "I am looked upon as the leader of a certain school of psychological theories," he bragged to his parents. "There are two or three camps among the faculty and graduate students, according to the special system of the science which each follows. The Behaviorists, whom I represent, have acquired a good deal of strength."[72] Indeed, Skinner believed that the director of the psychological laboratory and the most powerful figure in psychology at Harvard, Edwin G. Boring, considered Keller and him "serious threats" to the more traditional psychologies.[73]

He did more in graduate school than tinker, invent, and discover, however. He also wrote a theoretical, groundbreaking dissertation, which raised Boring's eyebrows. Boring was a daunting figure who had studied psychology at Cornell under another formidable psychologist, the eccentric, English-born, uncompromising, introspective structuralist Edward Bradford Titchener. In 1922 Harvard offered Boring a position as assistant professor, which he accepted, resigning from a full professorship with less pay at Clark University. He was hotly criticized during his first six years at Harvard, especially with regard to boosting psychology in relation to philosophy.[74]

When Skinner came to Harvard, Boring had just completed *A History of Experimental Psychology* (1929), a work that reverentially traced the past of the science back to its founders with eclectic thoroughness. Sometime in the spring of 1930 Fred wrote a review of the book for *The Saturday Review of Literature* in order, he told his parents, to help sales of Boring's book and help his own standing in the department.[75] Later he informed them that Boring was honored to have the review appear.[76] His reaction to Fred's dissertation on "The Concept of the Reflex," however, was another matter.

Graduate students remembered Boring as a good-humored but intimidating workaholic who kept a timer ticking on his desk to limit conversations to five minutes—to the second.[77] Boring served as dissertation adviser for another student who recollected anticipating their sessions with dread:

> Not only was Boring extremely critical from the technical and theoretical standpoint, but sitting there in the study of the Chairman of the Division, with books and journals lining the walls to the ceiling, you always had the feeling of pitting your own vast ignorance against Boring's vast Titchenerian knowledge. He seemed ready at the drop of a question to jump up and in his precipitate manner, clamber up the ladder, pounce on a particular volume on an upper shelf to find the exact reference where somebody . . . had already dealt with the problem you had posed.[78]

Although not a Titchenerian who introspectively studied mental "elements," Boring had little sympathy for the latest American psychologies—gestalt and behaviorism—which were then in competition with the functional and clinical approaches. Keller remembered Boring asking him, "when you wake up in the middle of the night out of a dream, do you believe psychology is the study of mind or is the study of behavior?"[79] This rhetorical question revealed Boring's limited understanding of Skinner's and Keller's field of interest.

In October 1930 Skinner received a five-page, typed criticism of his dissertation from Boring. Although Boring praised the writing style and recognized the novelty, he was not pleased. Later, to his credit, he removed himself from Skinner's dissertation committee rather than block approval and thus threaten the professional future of this up-and-coming and, indeed, rather uppity graduate student.[80] But at the same time Boring disapproved of the work, others, including Crozier, praised it. Moreover, a slightly revised version of the thesis would soon be accepted for publication in the *General Journal of Psychology*.[81] If Boring had continued to disapprove, he might have risked tarnishing his own professional reputation.

At first the dissertation appeared to be a history of the concept of the reflex, from René Descartes's argument that the reflex was mechanical action that explained all involuntary movement in animals and humans, to the experimental work of late-nineteenth and early-twentieth-century physiologists, such as Charles Sherrington and Rudolf Magnus, who showed with precision that the spinal cord controlled the action of nerves and muscles. Skinner had seemingly written a historical treatise on the advances of physiological science.*

*From Skinner's perspective, however, this history had been "unfortunate," for it relegated understanding "of the reflex as a form of movement unconscious, involuntary, and unlearned." This, in

What offended Boring was Skinner's use of history to support his novel argument. "I fear that you may be distorting history," he wrote to Skinner, a criticism he had of Titchener as well: "His method was to make a fine show of consulting all the authorities . . . killing off those that did not suit him and then seeming to find just what he always had been wanting." Boring accused Skinner of using the work of physiologists who studied neurone chains to show that they were really only finding stimulus-response correlations. This was only Skinner's interpretation, but he made it appear as if it were fact.[83]

Skinner's dissertation clearly challenged historical definitions of the reflex. Those definitions had progressively narrowed the utility of the concept, while they simultaneously resorted to quasi-metaphysical concepts such as the "reflex arc" and "synapse" to explain specific or "molecular" nerve action.* Instead, Fred boldly suggested that the concept of the reflex be given a broad, operational definition, that it describe the functional behavior of the whole organism. The reflex was simply to be considered a correlation between stimulus and response and other variables *outside the organism.*† Boring objected that "reflexology seems to be for you all of physiologized deterministic psychology, which is precious close to saying that it is all of psychology." He added that with this new, broader definition of *reflex,* he thought Skinner was being "a little disingenuous, that you were making a controversial argument under the guise of factual description—as Titchener used to do."[86]

Skinner recollected that he had always been interested in "getting at the dimensions of psychological terms and entities."[87] And, indeed, he was never merely an experimentalist content with the laboratory. From the beginning of his scientific career, Skinner loved to think and write about the broad theoretical terrain of the philosophy of science. His expansive outlook got him in trouble with Boring and would later, in more popular

---

turn, meant that "volition . . . was essentially the hypothetical antecedent of movement . . . which served to identify the reflex with scientific necessity and volition with unpredictability." He was interested in defining the reflex only in operational or functional terms and felt that the physiologists had made an unreal distinction between the involuntary (reflex) and the voluntary (volition).[82]

*He argued, for example, that "the synapse . . . described in terms of its characteristics, is a construct." There was simply a more scientific way of describing such phenomena. "There is nothing in the physiology of the reflex which calls in question the nature of the reflex as a correlation, because there is nothing to be found there that has any significance beyond a description of the conditions of a correlation."[84]

†What was outside the organism was a third variable: "The question of third variables is of extreme importance in the description of the behavior of intact organisms."[85]

forms, bring a firestorm of academic criticism. It is well to underscore the magnitude of his early intellectual ambition, for it was often his larger applications of behavioral science that both attracted adherents and repelled contemporaries. Skinner's rat experiments themselves were not at issue; rather, it was Skinner's radical theoretical departure from all previous definitions of the reflex. Abandoning conventional thinking about the reflex was an essential part of his science, but this breadth of vision, his speculative boldness, also signified that he expected and wanted more from science than did most of his colleagues.

Boring surmised that Skinner's new definition of the reflex had no place for spontaneous activity: "If there is any freedom in behavior, then to that extent the reflex is not all of behavior; but the determinist will not think that there should ultimately be any freedom left."[88] Boring was not convinced, as Skinner seemed to be, that all behavior was determined.

But something other than Fred's selective use of history and encompassing determinism was also at issue. He had violated the conventional role of a graduate student—that of a bright but deferential underling whose research adds to an already established tradition. Skinner had charged like a psychological conquistador into unexplored scientific territory. "You need more than a paper," to bring about such a bold change, Boring lectured; "you need propaganda and a school." He predicted that "Skinner's school of reflexism" would end like Köhler's Gestalt psychology, all semantics and no substance.[89]

Skinner was not at all put off by Boring's prophecy. He underlined *You need propaganda and a school* on the letter critiquing his dissertation and wrote in the margin, "I accept the challenge."[90] And in a formal reply to Boring, he wrote that Boring had read the thesis with the preconceived notion that it was a polemic for behaviorism. He pointed out that the four other readers had approved the thesis, and ended his note with a sarcastic version of Robert Hood's poem "The Bridge of Sighs," whose subject is suicide. The fate of the thesis was "up to the committee. To them I shall submit it—Owing its weakness. / Its evil behavior. / And leaving with meekness / Its sins to its Savior."[91] Years later Skinner was amazed that Boring let such irreverence pass.[92]

In early December 1930, with the dissertation still not approved, Boring wrote a letter of fatherly advice, praising and chastising, calculated to bring a promising but misguided psychological son back into the fold:

> I do not mean to be harsh, but your very versatility and you[r] polemical cleverness make it necessary for some older people to tell you bluntly where they think the trouble lies. Otherwise you might go on through life doing half-baked work which wins applause from the uncritical and the unsophisticated, working hard and sincerely, and thus never realizing that your work was superficial. . . . You have very unusual experimental ability; you have exceptional drive; you write well; your enthusiastic personality will make you a stimulus to others; you think clearly when your drive does not carry you away. The only flaw in this gem is that [you are] too clever always to be thorough . . . [and you] believe . . . that tricky sophistical argument is justified if the end is justified. Am I just old-fashioned?[93]

Boring was obviously ambivalent. Here was a student with the creative brilliance he himself lacked. But Skinner's lack of scientific deference rankled the older man's vanity and consigned all schools of psychology except behaviorism to scientific irrelevance. He felt a duty to promote Skinner as well as to protect tradition and what he believed was scientific propriety. In a general sense Boring faced the same problem with Skinner as Skinner's father had: how to give scope to the younger man's talent and at the same time make sure he did not reject what they saw as the real world of more conventional opinions, whether of businessmen or psychologists—in short, their worlds. But Skinner had discovered a new experimental technology with remarkable predictive power. He had charted the theoretical outlines of a radically new way of thinking about reflexes. In fact, he was now living in a new scientific world that he himself had shaped. He would not be turned back.

HOW HAD SKINNER come to such a novel and provocative theory of the reflex? He had read Pavlov's *Conditioned Reflexes* (1927) in Greenwich Village, and he had purchased Charles Sherrington's *Integrative Action of the Nervous System* (1906) before the end of his first semester at Harvard. But his most intensive reading in the history of the reflex occurred in late 1929 and early 1930 at the Harvard Medical Library in Boston. There, where he was often the only researcher, he delved into the original works of René Descartes, Robert Whytt, Marshall Hall, and Rudolf Magnus. Skinner remembered with pleasure the hours he spent totally absorbed in these works: "I'd go over there . . . and all these first editions were there and the librarian would get books for me so I had Descartes and Whytt. . . . I loved all of that." The reading in the medical library "kept bringing back this simple theme that all they observed was the correlation of stimulus and response."[94]

But Skinner remembered adding something different in his correlative formula: "I put the variables outside [the organism]."[95] If the variables were outside, they were environmental; and focusing on environmental variables was vital to the emergence of his science because it meant being concerned with the *behavior* of the organism rather than with the organism itself. Ultimately, he would not be concerned with motivation or drive or, in the end, even with learning. He would not study one thing in order to infer something about something else. He would not study physiology, or muscle reflexes, in order to study behavior. Animal behaviorists such as Thorndike and Hunter had dealt with the behavior of the entire animal, but no one had presented this experimental perspective in terms of a radical departure from existing literature on the reflex. Skinner, like Pavlov, was still talking about stimulus and response, which he would not abandon until 1935.[96] But in 1931, he was not completely Pavlovian because he was interested in the behavior of the whole organism. Moreover, he and Pavlov "were studying very different processes" once Skinner stumbled onto intermittent reinforcement: "Pavlov found it very hard to sustain salivation if food was not always paired with the conditioned stimulus, but rats pressed a lever rapidly and for long periods of time even though reinforcement was infrequent."[97] Skinner readily acknowledged borrowing the term *reinforcement* from Pavlov:

> I got the word from Pavlov and feel that it has a distinct advantage over "reward" by identifying the effect of a consequence of behavior in strengthening the behavior—that is, in making the behavior more likely to occur again. The old idea of pleasure and pain and Thorndike's adjectives "satisfying" and "annoying" refer to feelings which, in my point of view, is quite off the track.[98]

Skinner's behaviorism had other intellectual influences too. Recall that he had been impressed with Bertrand Russell's account of Watson's behaviorism during his Dark Year in Scranton. And his friend Cuthbert Daniel introduced him in the summer of 1929 to the Harvard professor Percy Bridgman's operational science in *The Logic of Modern Physics* (1927), in which Bridgman dismissed all definitions of science that were not based on observables.[99] Bridgman avoided complicated intellectual formulas, which obscured rather than facilitated real science. He asked simple but telling questions such as, "what is force [other] than certain kinds of observations in which you get rid of the thing?" That insight complemented Russell's position that "the concept of the reflex was comparable to the concept of force in physics." Science was not about specific nerve reactions called synapses or reflex arcs; it was about "fields of force."[100]

Skinner was also greatly impressed by the German physicist/physiologist Ernst Mach, to whom he was introduced in George Sarton's history of science course. He recalled being "bowled over" by Mach's *Science of Mechanics* (1883; first English edition, 1893).[101] Mach's approach to science owed much to Darwin: Science was simply a practical tool that had evolved to make necessary adjustments for survival. Science controlled facets of the environment in order to solve human problems, which, in turn, promoted the survival of the species. Science and technology were the same in that they were both problem-solving tools. Mach's model of science was parsimonious: Do not infer unseen things when you cannot analyze nonobservables.[102] Both Bridgman and Mach sought to rid science of all metaphysical and religious assumptions, stripping it down to the description of phenomena perceived by the senses. Science should not explain, but simply describe. Fred gave his nod of approval to both Bridgman and Mach in the introductory paragraph of his dissertation.

Another influence had been Jacques Loeb, who had corresponded with Mach and had studied the movement of the entire organism. Loeb had achieved considerable notoriety from his experiments in artificial parthenogenesis, which convinced some that he had discovered "the secret of life." Loeb believed that the object of science was not simply to observe phenomena but to control them. Skinner had read Loeb's work back at Hamilton, but with little immediate scientific effect. Crozier, who had studied with Loeb, also underscored the importance of making a scientific distinction between studying the action of glands inside the organism—as Pavlov had done—and studying the intact organism—as Loeb had done. Science was, Loeb said, "the study of forced movement [of the whole] organism."[103]

A final influence had nothing to do with intellectual inspiration. Skinner believed his own behavior as a scientist was determined by the nature of the experiments he devised and conditioned by his experimental subjects. He recalled a cartoon in the student newspaper *The Columbia Jester* of a rat saying, "Gee, have I got this fellow conditioned. Every time I press the bar he drops in a piece of food." Skinner said "it has often been argued that somehow or other this exposes the whole thing; it's a joke. But this isn't a joke. [I] have been conditioned by the organisms [I've] studied."[104]

The behavior of the scientist was not only determined; it was determined by the behavior of what was studied. The behavioral scientist was not some neutral observer who stood outside the experimental world and observed, unaffected. The experiment shaped one's behavior as surely as it shaped the subject's. Here was an implication that would eventually have a profound influence on Skinner's belief that his behavioral engineering applied to humans just as surely as it did to lower organisms.

He wrote to Saunders in January 1931 that he expected "a good deal of trouble" in defending his dissertation, as he had openly opposed the department. "But the degree becomes less and less important." He signed the letter, "As ever (but infinitely more happy)."[105]

By March he had completed all the requirements for his doctoral degree, but he did not immediately leave Harvard. With Crozier's aid and Boring's support, he applied for and received fellowships that would keep him at Harvard to continue his experiments in the laboratory of general physiology until the summer of 1936. He obtained a National Research Council Fellowship for 1931–32, and was reappointed for 1932–33. The 1932 appointment provided him with a stipend of $1,800 and stipulated that he would work with Crozier at Harvard and also at the Harvard Medical School with Professors Hallowell Davis and Alexander Forbes.[106] In a letter supporting Skinner's application, Boring noted that his interest in psychology was "rather limited to the problems of animal behavior" but that he was "extremely able and very original. . . . He has a feverish industry and an ebullient enthusiasm. We think of him here as having possibilities of 'genius,' using that word to represent this picture." But this picture was not without "one serious defect—a desire to found a school or a science of behavior upon research of the type which he is doing. Epistemological work of this sort is not within his sphere of training. I hope he can have the fellowship because it will direct his energies away from this kind of enthusiasm."[107]

On March 20, 1933, his twenty-ninth birthday, Skinner was interviewed by the newly formed Harvard Society of Fellows and awarded the following month the prestigious Junior Fellowship. A newspaper article proclaimed the fellows as "Harvard's new 'aristocracy of brains.' " Burrhus F. Skinner was considered "more reserved" than other fellows, "but a 'regular' fellow" anyway. As a member of the Society of Fellows he received a yearly stipend of $1,250, free room and board, and the use of the facilities of Harvard University.[108] He wrote to his parents that "it was probably the greatest event in my life" and went on to describe the exclusive club and dining room in loving detail:

> We sat down thirteen at dinner. On the table were thirteen solid gold candle sticks with the name engraved on each as place cards. I will get my candle stick when I leave the Society. . . . We had probably the best dinner I have ever eaten—sherry and bitters before, imported Dutch beer and cognac and other liqueurs afterwards. Green turtle soup and steak 3 ″ thick. . . . The conversation was wonderful and we were at the table more than three hours. Very genial atmosphere and very brilliant. The other Junior Fellows are fine, although I have not got to know them well yet. . . . It was all very wonderful and I was genuinely thrilled.[109]

William and Grace must have been bursting with pride—and no doubt a bit envious as well. Their Fred had arrived, and the tone of his letter shows that he wanted his parents to know it.

The new membership enabled Skinner to continue his research with no teaching duties. He would later admit that he was completely absorbed in his experiments. The luxury of more free time to indulge his interests and prepare and publish articles made this period of his life seem almost idyllic. Moreover, like many other young American intellectuals during the early 1930s, Skinner had found a cause. But his cause had nothing to do with the rush to embrace socialism or communism as capitalism seemed about to collapse; it had nothing to do with political radicalism or even support for Franklin Roosevelt's New Deal: "I made a quite explicit decision to devote myself *entirely* to my work. I did not read newspapers. I did not inform myself about elections or vote. And I was subject to criticism. The depression was at its worst, and there I was studying how white rats press levers."[110] Not only was he totally absorbed in his work but he was "out of step" with the rest of his profession. "In the thirties I seemed to have taken up rat-psychology just as it was dying [and] when social psychology was aborning. I was fighting for outmoded causes—Watsonian behaviorism, 19th century optimisms."[111] He complained to Crozier that other psychologists were wasting their time. "Why do you object?" Crozier replied. "There is just so much more left for you to do."[112]

Skinner was about to discover that Crozier's insight applied not only to his scientific career but to other facets of his life as well. His life up to 1936 had been relatively secluded, although it had fostered extraordinary qualities and remarkable discoveries. But his isolation was about to end as he entered a new world of personal, social, and experimental challenges. There was indeed much more left for him to do.

5

Behaviorist at Large

The attempt to force behavior into the simple stimulus-response formula has delayed the adequate treatment of that large part of behavior which cannot be shown to be under the control of eliciting stimuli. It will be highly important to recognize the existence of this separate field in the present work.

—B. F. Skinner, *The Behavior of Organisms: An Experimental Analysis* (1938)

I was humbled when I went to Minnesota and by Eve in my marriage.

—B. F. Skinner, Basement Archives (1963)

If one looked only at Skinner's experimental and intellectual life in the early 1930s, it might appear that his successes, in the security of the Harvard environment and with the luxury of several fellowships, were automatic—building effortlessly upon his discoveries in 1930. In truth, the rise of Skinner's science was fraught with problems and filled with more discoveries. And his personal life was to undergo a radical change.

It was perhaps Skinner's greatest fortune to experience marriage, teaching, and parenting only *after* his efforts to create a science of behavior were well advanced. The timing of major life events was one variable Skinner could not control. Throughout his life, he affirmed the role of happenstance: "It is amazing the number of trivial accidents which have made a difference. . . . I don't believe my life was planned at any point."[1] He did, however, take stock at various times in his life to set up professional goals, as mentioned earlier.

As the country battled the Great Depression, B. F. Skinner toiled in relative obscurity compared to Yale's Clark L. Hull and the University of California's Edward Chace Tolman. Hull was engaged in figuring mathematical equivalents for laws of behavior, while Tolman was analyzing "purposive behavior." Moreover, both were professors with loyal graduate students and supportive behaviorist colleagues.[2] In the decade to follow, Skinner would put together an experimental program that would eventually doom Tolman and Hull to relative obscurity while his own professional star rose.

As he established a scientific reputation, he found himself breaking away from other behaviorists—Hull and Tolman as well as Pavlov. There was a radical strain in Skinner's science that departed from all the others and that would, in the end, mark him as the most scientifically formidable as well as the most socially inventive of all the twentieth-century behaviorists. Skinner felt uncomfortable not only with mentalist psychologists; some behaviorists also made him ill at ease. While he was a Junior Fellow, he tried especially to distinguish his experimental approach from Pavlov's stimulus-response psychology:

> I simply do not believe that S-R theory has enough to offer a science of behavior to make it worthwhile. I do not believe that experiments are best designed . . . through any deductive process. We can learn more about the behavior of organisms by direct observation of their behavior, particularly frequency of response, and by watching this variable change as a function of independent variables in which we may be interested.[3]

Although he had discovered in 1930 that immediate reinforcement resulted in a smooth ingestion curve—that is, that a rat could be instantaneously conditioned to a lawful rate of response—he still faced a major difficulty. There were obvious variations in the animal's behavior while in the experimental box. The rat, for example, made any of several different movements before pressing the lever. What determined these varying behaviors? What specific stimuli produced exploration of the cage, standing on hind legs, moving forelegs, and the first pressing of the lever? The history of reflexology in general and Pavlov in particular demanded that a specific stimulus be linked to a specific response—for Pavlov, the sound of a bell to the flow of saliva. Here again arose the problem of variability he had attempted, with some success, to eliminate as he devised his apparatus in 1930. Although certain that immediate reinforcement produced lawful behavior, Skinner could not adequately account (in terms of stimulus-response reflexology) for other movements the rat made in the enclosed experimental environment. Rather than attempting to locate specific stimuli (such as what made the rat stand on its hind legs), he admitted that some of the rat's behavior appeared to be spontaneous.

Skinner's solution to the rat's variable behavior was, to a large degree, to ignore it, concentrating instead on something he was sure of: the lawful *effects* of the delivery of reinforcers to an intact organism in an enclosed space. This decision became the key to Skinner's experimental success after 1930. Behavior was conditioned (or "shaped") by its consequences. The reflex, as a direct link between a stimulus and a response, was no longer the center of Skinner's attention, as it had remained for Pavlov.

Skinner's original insistence that virtually all behavior was stimulus-response (the position he took in his dissertation) had become by 1936 an assertion that some behavior must be assumed spontaneous until an identifiable stimulus could be found. The dimensions of the study of behavior were radically altered. It was, for example, no longer necessary for an experimenter to concentrate on the crucial role of a stimulus—such as Pavlov's bell—in eliciting a conditioned response, in that case the flow of saliva. The "emitted" behavior itself would be studied, a shift of attention toward the function of the contingencies that conditioned or effected a rate of response. The rate of response, or rate of lever pressing, was the behavior that could be controlled by changing experimental contingencies, such as the state of food deprivation or a particular schedule of reinforcement.

But Skinner's statement that some behavior is "spontaneous" until a specific eliciting stimulus is found moved him dangerously close to indeterminism and the position he had resisted from the beginning, namely, that the organism has free will and can choose how to behave. He needed to justify theoretically his emphasis on the effects of reinforcement as lawfully determined behavior, and somehow contain the variability-of-behavior problem he had frankly admitted. By the spring of 1934 he was hard at work on an essay that would discuss the "generic nature" not only of the reflex but of the lever-pressing behavior as well.[4] Some behavior might *appear* to be undetermined or spontaneous, he explained, but certain "classes" of behavior can be brought under the control of reinforcers to show orderly, repeatable rates of response. Lever pressing was a class of behavior made up of a number of separate movements for which no identifiable stimuli could be found. Nonetheless, one could get lawful rates of lever-pressing responses by changing schedules of reinforcements. And these rates of response could be obtained again and again and were predictable under different schedules of reinforcement. That was as much behavioral science as reflexology.[5]

In 1937 Skinner called lever pressing an "operant" class of behavior, since it was an operation (a behavior) emitted without any readily identifiable eliciting stimuli.[6] To Skinner's continuing amazement, when an unelicited press and its attending noise were immediately followed by the delivery of a pellet in the food tray, the rat continued to press the lever at a rate of response that translated as a smooth curve on the cumulative recorder. The operant (the rat's lever-pressing behavior) thus came under the control of a reinforcer and was conditioned.

Skinner recorded this new kind of lawful behavior in dozens of experiments between 1932 and 1936 and described them in some twenty published articles. He showed, for example, in "Drive and Reflex Strength," "Formation of a Conditioned Reflex," "Extinction of a Conditioned Reflex," and "Discrimination," that rate of response or lever pressing was determined by the manipulation of variables in the experimental box.[7] Such behaviors were depicted on ever more sophisticated curves of cumulative records. Skinner focused on the reinforcement of a *behavior* in strengthening more of the same kind of behavior, rather than on a physiological reflex response—say, a frog's muscle twitch after an experimenter touched an electric current to the frog's leg muscle, or the salivation of Pavlov's dogs when a bell was rung.

This was indeed a new experimental science of behavior, a point initially missed even by Skinner's most loyal behaviorist friend, Fred Keller. Keller admitted that Skinner's experimental novelty had initially escaped him, but he caught on when he read *The Behavior of Organisms.*[8] Skinner published that book in 1938 as the culmination of both his experimentation and his theoretical development since 1930.[9] His lever-pressing operant was a unit or class of behavior that could be studied just as generically and scientifically as the Pavlovian reflex.

Skinner's operant science emerged when reflexology was less credible than it had been earlier in the century. Watson had suggested that reflexes were linked or associated and that they determined the complex behavior traditionally called "thinking." Physiologists like Sherrington and Magnus were skeptical of applying the reflex to complex mental life. Any advance in using conditioning to explain complicated behavior had to come through correlating stimulus with response in the laboratory. By the time Skinner discovered the regularity of the unelicited response, Watsonian conditioning was being dismissed as crudely mechanistic, going far beyond the laboratory and, hence, scientific credulity.[10] In a general sense, Skinner had put Watson's failed promise of a science of behavior back on track. And he challenged the Pavlovians by arguing in 1935 that his *operant* conditioning was a second type of conditioning, Pavlov's *respondent* conditioning being the first.*

One can, however, go too far in emphasizing differences between Pavlov and Skinner. After all, they both used reinforcers and got conditioned, or determined, responses. Moreover, Skinner did not abandon the stimulus as an experimental variable, even though many stimuli were not readily identifiable under experimental conditions. He was, for example, interested in how rats responded to a "discriminative stimulus"—when, for instance, food was delivered to a rat only when a light was on in the box.[12] But it was the *effect* on rate of response of a new variable that concerned him more than the stimulus or its connection to a specific physiological response. Skinner, like Pavlov, also emphasized that the organism did not "choose" among different stimuli; rather, its behavior was determined by the conditioning. Mental processes did not determine either respondent or operant conditioning.

THE BEHAVIOR OF ORGANISMS received a largely negative response from American psychologists. (Skinner later remarked that "a purely descriptive science is never popular.")[13] Reviewers criticized his failure to relate his operant system to the work of contemporaries and their predecessors. The Stanford psychologist Ernest R. "Jack" Hilgard took issue with the accuracy of the book's title, since Skinner had experimented only with white rats. And although Skinner had made an important distinction between two types of conditioning, he had not explained the relationship between his system and those of other behaviorists.

*Skinner never directly engaged Pavlov in an argument about the merits of operant versus respondent conditioning. He was, however, challenged by two Pavlovian Polish reflexologists, J. Konorski and S. Miller, who disagreed that lever pressing was really strengthened, that is, that the rate of response increased by reinforcement. They argued that the strengthened behavior was simply the formation of a new reflex.[11]

Hilgard asked, "If Skinner has been unable to relate his work to that of other investigators, how can a reader, coming fresh upon this new body of materials, be asked to make the transition?"[14]

Skinner's defense was that "the dimensions of operant behavior provide a new approach which is simply not to be found in any significant way in earlier research," and thus he was at a loss as to how to relate his work to that of others.[15] Years later he suggested a research project to substantiate the charge that he neglected current animal psychology: "It would be an interesting bit of historical research to go back and see just what it was that I had 'neglected.' I suspect that everyone neglects it now."[16] His book had merely provided "effective experimental methods and an appropriate theoretical formulation." And he concluded that "the difference between operant and non-operant research is, so far as I am concerned, almost entirely one of the dimensions of the thing studied." As he had been with his dissertation, Skinner was still "concerned with getting at the dimensions of psychological terms and entities."[17] To have continued to study reflexes would have narrowed his field of experimentation. Besides, his experimental results proved he had launched a science based on the study of behavior for itself alone.

Skinner did receive some praise and encouragement, however. Edward Chace Tolman wrote to him: "I haven't been so intellectually excited in a long time as I am now perusing your book. It is, of course, a very major contribution to 'real' psychology." He suggested that Skinner experiment with two levers, as Clark Hull was presently doing.[18] Walter Hunter congratulated him on his "very nice treatise on behavior. It is a decided achievement and should be a great professional asset."[19] And Keller, then teaching at Columbia University, surprised Skinner by incorporating the book into his lectures. "You had more faith in that book than I did, by a heck of a lot," Skinner recalled. "You started teaching from it and I didn't do that for years." Keller had "always been doing revolutionary things in teaching."[20]

The most radical implication of this new behavioral science was its potential social application. Skinner never doubted the transferability of operant conditioning from white rats to other organisms, including human beings: "The importance of a science of behavior derives largely from the possibility of an eventual extension to human affairs." The only difference he anticipated aside from the vast complexity of higher organisms was in verbal behavior.[21] Indeed, the experimental results reported in *The Behavior of Organisms* would eventually influence the development of teaching machines, programmed instruction, and community treatment programs for juvenile delinquents. Conditioned reinforcers were adopted in industry to prevent on-the-job accidents and to help mentally retarded and autistic children, as well as in mental health management, physical rehabilitation, behavioral medicine, chemical dependency, pediatrics, and many other areas.[22] Reinforcement principles were also put to use in airline "frequent-flyer" programs and certain credit-card promotions. Other applications were made in training monkeys to assist paraplegics and in training porpoises and whales for popular entertainment. Various branches of the military also used operant techniques and positive reinforcement.

But Skinner would not begin to apply the social potential of his science until the 1940s, and correspondence with other American behaviorists indicated that he was having trouble in the mid to late 1930s in bringing them over to his purely descriptive or radical behaviorism. "I wish to hell you would get interested in this side of the matter," he told Keller. "You can control your drive [the rat's state of food deprivation] quite well enough to rest easy about it, and there's a hell of a lot more fun and a better chance for results that will make you famous."[23] Moreover, even though the most prominent American behaviorists, Tolman and Hull, were impressed with Skinner, neither ever abandoned his own behavioral perspective for Skinner's more wholly descriptive and less deductive experimental approach.

Tolman, a Harvard doctorate, had worked with Hugo Münsterberg, a pioneer in applied psychology, and for a brief time had flirted with Gestalt psychology. Tolman's major book, *Purposive Behavior in Animals and Men* (1932), had "doggedly worried the last drop of generalization from his laboratory studies and . . . the rat [emerged] with a psychological dignity which no other behaviorist [had] granted even to the ape."[24] Skinner had met Tolman in the summer of 1931 when the latter had returned to Harvard to teach a psychology course. They talked frequently on such topics as drive and reflex strength, and Skinner's dissertation idea of a "third variable." Skinner was miffed that Tolman used the idea of the third variable without acknowledging him in a famous paper Tolman published three or four years later. He called it instead an "intervening" variable, which Skinner took to be "the great mistake of all cognitive psychologists."[25] Skinner placed the third variable outside the organism, while Tolman and others substituted an intervening variable for a mental event.

Hull wrote Skinner in 1934 that he had been "following your work with interest. . . . [It] fits in very well with the similar approach which is being developed by myself and a number of my associates here at Yale."[26] He invited Skinner to speak to his graduate students in New Haven, and it was Hull who deserved the credit (or blame) for first referring to the lever box or operant chamber as a Skinner box.[27] Skinner respected Hull as a scientist, even though he felt that his attempt to fashion deductive postulates for a generalized science of behavior was destined to fail and that the future of any viable behavioral science lay on the experimental side, especially with the development and application of operant conditioning.* An early letter confirmed Skinner's respect as well as their scientific differences:

> You are one of the only psychologists I know who appreciates the importance of examining the fundamental nature of a science of behavior and who [is] definitely working out its structure. That is also my chief concern and the only difference I can see between our approaches is the selection of the terms. . . . You regard science as the confirmation of hypotheses, while I consider it to be primarily simply descriptive.[29]

Skinner's letter puzzled Hull. While he had "the fullest appreciation of the value of systematic exploration and the testing of isolated hypotheses," which he identified as Skinner's research, he also maintained that their differences were "essentially a matter of taste, and surely there is no point in quarreling over what we like and what we do not like." He recommended that "for the good of your soul" Skinner should read Isaac Newton's *Principia.* Hull believed there was no insuperable problem in applying the theoretical ground of one science to another, and believed he could be the Newton of behavioral science. He admitted, however, a difficulty: "We do not . . . know exactly what to look for."[30] But Skinner was convinced he not only knew what he was looking for

*In a review Skinner later dismissed the scientific utility of Hull's most ambitious book. But he also recalled that Hull "used to explain my refusal to theorize as a fear of being wrong. I think there are other, and better, reasons for not theorizing, but he was right—I am afraid of being wrong." One behaviorist has recently commented that "looking back [at Hull], the image that often occurs to me is that of Icarus. . . . He flew too high, the sun melted the wax, and he fell into the sea." Skinner, however, avoided high-flying theory and "lengthy verbal linkages between theory and data" and kept "his concepts precise, simple, and relatively few in number."[28]

but had found it: a purely descriptive experimental behavioral science that was freed from physiology as well as mentalism.

Not surprisingly, his position was criticized by more conventional American psychologists. "Skinner is wrong," Brown University's Leonard Carmichael wrote to Clark University's Carl Murchison,

> in setting off behavior as something that can be studied without reference to physiological processes. . . . His own brilliant experimental work demonstrates that his thesis is to a degree sound in that some . . . problems can for a time be attacked most profitably in terms of his system. His view of the independence of a science of behavior illustrates to me not that behavior can be ultimately divorced from physiology, but that the two systems should be considered in terms of functional relationships and not, of course, as a naive mixture as exemplified by Pavlov.[31]

But Skinner was single-mindedly determined to pursue his science of behavior, even if it meant wholesale abandonment of all traditional psychological schools and studies.[32] Boring's earlier warning that his radical departure and theoretical cleverness would get him into trouble was being borne out even as Skinner paid it no heed. If anything, as experimental results and theoretical maneuverings made it clearer that he was on the right scientific track, he became even more aggressive. To S. S. "Smitty" Stevens, a young Harvard psychologist, who to some degree gravitated toward behavioral science, Skinner urged: "Let's get busy with the real job—the establishment of a system of behavior with respect to which all these things can be stated and dealt with experimentally. . . . Why not get out of that historical psychological bog known as the field of sensation?"[33]

To his dismay, few experimentalists, behaviorists or otherwise, jumped on the operant conditioning bandwagon in the mid-1930s. But although Skinner wrote that "I lacked encouragement. . . . all I remember is the silence,"[34] in fact he received praise from Karl Lashley, his former professor Walter Hunter, Leonard Carmichael, and William Crozier. The latter allowed his former pupil use of the laboratory and facilitated his publication in the Rockefeller Institute's *Journal of General Physiology,* of which he was co-editor. Skinner also received invitations to speak at Yale, Clark, Brown, and his alma mater, Hamilton. He attended the Mach Colloquium at Brown on experimental studies in the definition of stimulus-response, and he had talks with Alfred North Whitehead on behaviorism. Nonetheless, his time as a Junior Fellow was coming to an end, and the depressed job market permitted little hope for a decent academic position.

A double-exposed picture of Skinner began to emerge. His colleagues and contemporaries saw him as a brilliant and determined contributor to important scientific research; but they also found him argumentative, fanatical, and intolerant of other approaches. A letter from Boring summed up the mixed impressions:

> The differences between you and me always were, I think, that you had a very strong cathexis for your ideas, seeming a fanatic to me, and that you were definitely aware of differences from yourself and constantly directing your invective against them, whereas I did not feel these differences mattered very greatly and thought that they probably lay within the . . . [spectrum] of human fallibility anyway: although, of course, when your invective was directed against me, I tended to reply crisply. [You did] not feel that a certain distribution and variation of values is a good thing among scientists.[35]

Skinner, for his part, for years saw himself as being unappreciated by the scientific community.[36] Indeed, it was not until the 1940s that his science began to recruit a corps of dedicated graduate students.

WHATEVER HIS PROFESSIONAL ISOLATION, Skinner's social life was fairly lively. In 1933 his parents bought him a Ford coupe. He had a clavichord made, having previously owned a small red piano. His absorbing interest in experimentation did not prohibit occasional dating. In late 1929 he wrote his parents that he was seeing Grace DeRoos, "the last name being Dutch, which is as close to Grace Burrhus as I could get." Grace was a graduate of Bryn Mawr, had received an M.A. from Harvard, and was interested in physiology. Their compatibility seemed remarkable: "We disagree on nothing except the theory of nervous conduction and the right amount of olive oil in salad dressing. We are almost exactly alike. We even have only one tooth missing apiece, and it's the same tooth. . . . right now I have known her only 193 hours and 10 minutes, so it's a bit soon to pop the question."[37] But by early 1930, he reported that marriage was no longer a possibility because "we are too much alike to get along together when we are with each other for any length of time."[38] Nonetheless they continued to date each other. A portrait of the kind of woman Skinner might settle down with began to emerge: attractive and intellectual.

In 1932 he courted Victoria Lincoln, a friend of Cuthbert Daniel, the mathematician who had introduced Skinner to Percy Bridgman, and his wife, Janet. "Vicky" was a divorcée with a young child. She would one day marry Victor Lowe, a man Skinner later remembered as "the dullest man I ever came across." He never understood why this "brilliant girl" became attached to Lowe. What remained unsaid was why she never became attached to him. Vicky was working on a novel, came from a prominent family, and seemed to have the intellectual and class background with which Skinner sympathized. She eventually published her novel and the "dull" Lowe, her second husband, published a book on Alfred North Whitehead.[39]

Mary Ann Chase lived on Monhegan Island off the Maine coast, where Skinner spent "a very passionate summer" in 1932. He had begun spending summers there with friends. He recalled being "very much in love with her, but it didn't work out." They maintained a friendship after "that marvelous summer," although he only saw her occasionally.[40] After his marriage Skinner bought a cottage on Monhegan, where for years he spent summers with his family.

He also courted Barbara Channing, a distant relative of the Unitarian intellectual William Ellery Channing. "I liked her very much," he recalled. "But there was no affair."[41] She was apparently fond of him and composed poems for him. Skinner was attracted to women with literary interests. He regularly socialized with Olivia Saunders and her friend Mary Louise White. Olivia was the daughter of his old Hamilton mentor, Percy Saunders. She and Mary Louise had an apartment on Brattle Street in Cambridge and were frequently visited by the young novelist James Agee. Skinner would occasionally spend long evenings discussing literature with these three friends. But these were strictly social and intellectual relationships.

It was different with Ruth Cook, whom Skinner called "Nedda" in *Particulars of My Life,* and who was his most intense love interest during the Harvard years. He met her in the winter of 1933–34. She had taken her undergraduate degree at Swarthmore and was a graduate student in anthropology, interested in American Indians.[42] Almost immediately Skinner fell wildly in love. He recalled, however, that their social circles were dissimilar and, unlike his other female companions, Ruth had no pronounced literary interests. But in the heat of their relationship it did not seem to matter, and for a short period the couple all but lived together. When the break came it stunned him. At dinner one evening, "she told me that we should break it off. She had been more or less engaged to a young man who was chronically ill, and she was going back to him."[43]

Skinner fell into a deep depression: "I was almost in physical pain, and one day I bent a wire in the shape of an N [for Nedda, or to protect Ruth's identity in his memoir], heated it in a Bunsen Burner, and branded my left arm. The brand remained clear for years." Indeed, it was still faintly visible at the end of his life. He recalled that his mother asked about the brand, and when he said it was a burn from an experiment, "she wisely let it go at that."[44]

Months afterward Ruth unexpectedly reappeared, showing Skinner a note from a physician stating that she was pregnant. She implied that he was the father and asked if he would pay for an abortion. He asked her to marry him, have the baby, and then divorce him and let him keep the child.[45] When Ruth refused this unusual offer, Skinner came up with $400 for the procedure, going so far as to contact a loan company in Boston, before friends lent him the money.

Skinner learned later that Ruth had been seeing yet another man after their affair had ended. Besides, he had always used contraceptives with her.[46] Beyond the question of fatherhood, however, the liaison and its unexpected consequence revealed the risks Skinner was willing to take for her. Ironically, it may have been Skinner's family who indirectly persuaded him to assume responsibility: "There were no verbalized rules of honor in my upbringing, but some things were right and proper and I did them when occasions arose. One helped a friend, one helped a former lover now in trouble."[47]

Skinner's longing for happiness with a woman, a longing that was dashed time and time again, contrasted sharply with his scientific successes. Later, he particularly enjoyed the music of Richard Wagner, whose operas resonate with powerful desires, smashed hopes, and ascendant yet long-suffering emotions—a musical mirror, perhaps, of his own feelings.

Skinner's personal suffering notwithstanding, he almost certainly avoided marrying the wrong person in Ruth Cook. Whether he eventually married the ideal person, even though his marriage lasted well over half a century, is open to question. His love interests were always tempered to some degree with ambivalence and disappointment. And, indeed, Skinner would have a reputation as a womanizer.

YVONNE BLUE (LATER "EVE") was the oldest daughter of a well-to-do ophthalmologist from Flossmoor, Illinois, an affluent Chicago suburb. She and her two sisters enjoyed a childhood of cultural advantages and the luxuries of an elegant house with a library and a tennis court. She was not required to do domestic chores, nor was she given any responsibility for looking after her younger siblings. Yvonne had a famous relative, her maternal grandfather, Opie Read, an American humorist who introduced the young Yvonne to the world of literature, to which she became enthralled.[48] At thirteen she confided to her diary her hopes for a literary career: "I am going to start as a newspaper reporter and work up. Maybe I'll get a position as a book reviewer or movie reviewer. Then I'll be on the editorial staff of a magazine. . . . And maybe someday, I'll write books!!"[49] She majored in literature at the University of Chicago, and took a course with Thornton Wilder there. She mixed with a bohemian crowd. To the horror of her parents (especially her mother), she smoked in public. They did not know about her sexual experimentation.

During Skinner's last year as a Junior Fellow, Yvonne visited a friend in Cambridge who introduced them on July 22, 1936. At the time, he recalled, she "was very much in revolt against her parents and their way of life."[50] Less than a year after their marriage, she considered him "the best person I've ever met, and I knew it the first five minutes I spent with him."[51] Here was an articulate, well-read, slim, handsome young man who enjoyed literary conversation over cocktails and gourmet food. He was also a Junior Fellow at Harvard, a passport to a prestigious future. Skinner found the tall, light-brown-haired Yvonne charming and attractive, and she could also get all of his literary allusions. They had similar upper-middle-class backgrounds and both enjoyed being "liberal hippies of our epoch."[52]

The day following their first date they visited Walden Pond and spent the evening in his rooms.[53] A few weeks later he persuaded her to come back to Cambridge, and introduced her to his parents, then vacationing at Cape Cod. They were pleased. Once the couple announced their engagement, however, Grace could not help mixing criticism with approval in a letter to her son: "We were very happy to receive your announcement and think you have done the right thing. We liked Yvonne very much and feel sure she will make a fine sensible wife. It will mean a great readjustment for you as you have had such a free independent life alone so long, but I guess you are equal to it."[54] The Skinners had worried that he might become a confirmed bachelor, so the announcement came as a relief. His style of living would become more like their own.

But before the engagement there was a glorious summer romance. They drove to Monhegan in his Ford roadster, skinny-dipped at night in the island's Gull Pond, and played chess on a set Skinner had built from mother-of-pearl buttons. In August he traveled to Flossmoor to meet Yvonne's parents and then continued on to Minneapolis, where he had accepted a teaching position at the University of Minnesota. He recalled being moved to marry Yvonne not only by their compatibility but also because he needed emotional support after leaving Harvard.[55] He urged her to visit him in Minneapolis so she could envision their life there.

But the visit did not go well. Yvonne was not enamored of academic life. The prospect of becoming a faculty wife did not thrill her; it would be a more conservative way of life than the bohemian style she had cultivated. And there seemed to be nothing in Minneapolis to compensate. It seemed provincial and dull compared to Chicago. Skinner remembered that the faculty couples Yvonne met did not impress her.[56] Suddenly she broke off the engagement, "because I knew he wasn't happy about it," according to her diary.[57] Skinner recalled, too, that he had had doubts, but he did not want to end the engagement.[58] He visited her in Chicago, and they made up in a hotel room. They were married in a justice-of-the-peace ceremony at the bride's home on November 1, 1937.

The Minneapolis years would be difficult ones, especially for Yvonne. Their social life revolved around other couples in the department, with whom she felt little kinship. By the time their first child, Julie, was born, in 1938, Yvonne had begun to adapt to Minneapolis, but it was a slow, difficult adjustment. She was not happy. She had no career and found the responsibilities of motherhood burdensome. (Their second child would be born in 1944.)

Initially, before the children, they lived in a one-room apartment with a kitchenette and Murphy bed. Yet she required a maid. The young couple still shared a love for literature, but their other interests diverged. Yvonne enjoyed dancing, something Fred was not very good at. He liked to attend symphony performances, while she found them tiresome. Their sleeping patterns were also at odds. Yvonne slept late and Fred was up at dawn, a routine that persisted throughout their marriage and meant that Fred tended to the early-morning needs of his young daughters.

As an old man, Skinner did not believe these idiosyncrasies had been the root of the difficulty: "She never gets anything out of doing things. . . . I don't think [anything] very much happened that reinforced the kinds of behavior that would have been part of a good marriage." Looking back over his very long marriage, Skinner concluded that "the thing that bothered me most . . . about Eve was that she never got the slightest satisfaction out of anything that I did. . . . I think Eve envied my success and didn't have one of her own."[59]

But was Yvonne really envious or simply indifferent? In a 1971 interview with *Cosmopolitan* magazine, she admitted that her husband "doesn't tell me much about his work because one, I'm not a psychologist and wouldn't understand, and two, I'm not terribly interested."[60] One could also ask whether he was really interested in the kind of life she enjoyed. He was deeply concerned about her, but that was not the same as wanting to become a part of her world—although they were both devoted to their children. Though she did not scorn her husband, as Grace Skinner had William, there was an interesting marital similarity there. Both Skinner and his father required considerable emotional support from their wives. Both were accused of supreme egotism, yet had fragile egos. Neither received enough adoration or support.

The war years certainly did not make things easier for Fred and Yvonne. To complicate matters, Yvonne's youngest sister, Norma Blue Calt (nicknamed "Tick"), whose husband, Ray, was in the service, lived for a time with the Skinners. Skinner was not drafted, but he did spend a lot of time away from home, working on a military project at General Mills Company. In 1945 he decided to take an offer to become chairman of the psychology department at Indiana University in Bloomington, a small, provincial town that Yvonne loathed.

The early years of their marriage were unsettling. Skinner candidly admitted having indulged in some "sexual experimentation" while in Indiana, which, he maintained, had little positive or negative effect on their relationship.[61] Once, after a crisis at the family cottage on Monhegan during the summer of 1949, she left him and the children and stayed for a short while in New York City.[62] He remembered the time between 1938 and 1953 as "the most troubled period of my life. . . . Interrupted by the war, two moves from place to place, [and] two and one half years of administration."[63]

So while marriage gave Skinner the security of a sustained relationship, which had previously eluded him, as well as the great joy of two daughters whom he adored, he remained to some degree a man with powerful, unresolved romantic feelings but determined to do what he could to build a workable, mutually satisfying relationship.

WHEN FRED FIRST MET YVONNE in 1936, he had no real prospects for securing a teaching position at a first-rate university. Aside from the shrinking budgets in many institutions because of the depression, there was another consideration. "Your reputation is already so great as to make them afraid of you," Boring wrote to him; "five years of uninterrupted research for a very able man makes him unplaceable in America."[64] Whether this was simply his mixed opinion of a bright but formidable graduate student or an accurate assessment is difficult to determine. While Boring remained ambivalent about Skinner, he actively supported him for several academic positions. Nonetheless, university departments often hired safe, conventional candidates over brilliant ones in order to avoid department controversies and jealousies.

Eventually Boring was responsible for recommending Skinner to his friend Richard M. Elliott, who chaired the psychology department at the University of Minnesota. Boring wrote Elliott, who was himself a Harvard doctorate, that "you have . . . your chance to get a young genius, who, under your beneficent protection, would blossom out even more than he has here."[65] Elliott replied that "your letter about Skinner turned the trick."[66]

Skinner may have had doubts about going to Minnesota. He was not keen on teaching and would have preferred a research position. Leonard Carmichael painted a positive picture of the move.[67] Walter Hunter also recommended that Skinner accept the position if it was offered. He was at Minnesota when he wrote: "The fellows here are splendid; the university is excellent; there are fine opportunities for research; and the teaching schedule would be light. Minnesota is a good place to be—either to stay or from which to get an offer. I hate to see you go so far from New England; but I really think it is a fine opportunity for you to make an academic start."[68]

Elliott offered Skinner an instructorship at an annual salary of $1,960.00. He would teach one class of elementary laboratory psychology and participate in colloquiums for Ph.D. students. He would, as Hunter predicted, have plenty of time and good facilities for research, and Elliott made sure the teaching duties would not interfere with his research.[69] Elliott's rosy picture of university affiliation helped convince Skinner to accept a one-year position, but Elliott had to fight the psychology faculty to bring him aboard.[70] Skinner later learned that he got the position only because Elliott regarded him as a "special, special case."[71]

Skinner entered an eight-member department, which, while certainly not unprofessional, was not especially distinguished. There was no one at Minnesota with a national reputation in psychology, although Robert Yerkes had been there briefly in 1917 and Karl Lashley had left in 1926. Elliott had achieved some recognition as editor of the Century Psychology Series (which would publish *The Behavior of Organisms*), but had done little original research. The department, like many in America, had long been associated with philosophy. That changed when Elliott took the chair in 1919. At that time graduate work toward the Ph.D. was offered, and when Skinner arrived over fifty students had received their psychology doctorates there. No dominant psychology held sway at Minnesota, and thesis work could be done in "practically any field": animal behavior, social and abnormal psychology, experimental psychology, individual differences, vocational psychology, and personnel.[72] Toward the end of the academic year, Skinner would write to Crozier: "My position here is all that could be desired. My teaching duties are light, my students interesting, and I have all the facilities for research that I need."[73]

The immediate professional effect of the move to Minnesota soon became apparent: "I came out of my shell. . . . It was in Minnesota where people asked me to talk about human behavior outside of the experimental laboratory."[74] Through his colleague William Heron he was exposed to other issues in psychology: "Heron was working on breeding bright and dull rats and the effects of drugs and it turned out that the operant technique was beautiful for assessing drugs."[75] He also received "strong reinforcement" for his teaching, a benefit that would continue for many years after leaving Minnesota, as he observed former students achieve positions of rank and importance in academic and political arenas.[76] The reinforcement Skinner received from his teaching was partially facilitated by the special arrangement Elliott made for the better students to be brought under Skinner's wing: "Partly through this device and in general through the appeal of his originality, Skinner quickly surrounded himself with a highly motivated group of students, many of whom were thus snared for advanced work."[77]

The eclectic nature of the department encouraged Skinner to pursue lines other than his rat work, most notably verbal behavior. His interest in literature had never waned. Yvonne read voraciously and introduced him to new novels. They often read Trollope aloud to each other. Earlier, as a Junior Fellow, Skinner had been challenged by the philosopher Alfred North Whitehead to explain language as behavior, and in 1934 Skinner began to collect examples of verbal performance that would eventually be used in a book on language.[78]

About the same time he wrote a short article for *Atlantic Monthly* on the automatic writing of Gertrude Stein.[79] And in 1935, while attending his rats, he suddenly found himself saying silently, "you'll never get out, you'll never get out, you'll never get out." He had unknowingly verbalized the rhythm of sounds from his rat apparatus! The experience prompted him to build a device he called a "verbal summator." A phonograph record played vowel sounds separated by glottal stops to a person who wrote down what he or she heard. The rhythm, just as with his rat apparatus, suggested real language. He began to collect samples of verbal responses.[80] He also befriended the linguist I. A. Richards, who was teaching at Harvard. Richards took a behavioral rather than a symbolic view of language and encouraged Skinner to broaden his reading.

While at Minnesota, Skinner became fascinated with alliteration, of which he found many examples in Swinburne's poetry but virtually none in Shakespeare's sonnets.[81] By the summer of 1937 he was offering a course on "The Psychology of Literature," which proved to be more psychoanalytic than behavioral but nonetheless kept him thinking about language. In 1941 he was granted a Guggenheim Fellowship to write a book on verbal behavior, but waited until 1944 to use it to write what became Harvard's 1947 William James Lectures and eventually *Verbal Behavior,* published in 1957.[82] His interest in verbal behavior gradually took priority over the work with rats, which continued but without any experimental breakthroughs.[83]

As it turned out, Skinner's work on verbal behavior was preempted by another interest, one directly related to the war that had burst upon Europe in September 1939 and sucked the United States into its vortex two years later. Indeed, if World War II had not intervened, Skinner might have published *Verbal Behavior* years sooner, before leaving Minnesota in 1945. The war provided him with a longed-for opportunity to move operant conditioning outside the lever box. He had the chance to apply operant conditioning to weapons systems—to try to improve missile guidance in a government-funded research program eventually called Project Pigeon.[84]

In April 1940 Skinner boarded a train for Chicago to attend a Midwestern Psychological Association meeting. En route he reflected on the horrors of airplane bombing and wondered whether some countervailing technology could be found to stop the bombers before they delivered their deadly cargo: "I was looking out the window as I speculated about these possibilities and I saw a flock of birds lifting and wheeling in formation as they flew alongside the train. Suddenly I saw them as 'devices' with excellent vision and extraordinary maneuverability. Could they not guide a missile? Was the answer to the problem waiting for me in my own backyard?"[85]

The idea was simply that birds might be trained as navigator-bombardiers. When he got back in Minneapolis, he found a poultry shop that sold pigeons to Chinese restaurants and bought a few. He found that if he restrained a pigeon's wings and feet (initially he slipped the bird into a toeless sock), the bird's sharp vision would allow it to peck a target, while its moving neck would produce steering signals. Using its head and neck, it could pick up grain as a reinforcer.[86] In the beginning he experimented not only with pigeons but with tame crows, which proved "remarkable subjects" but for their difficult temperaments.[87]

Norman Guttman, a student at the University of Minnesota who worked on Project Pigeon, described the messy environment in which the early project was conducted at the university:

> a picture of Stalingrad, much enlarged from *Life,* a balsa-wood model sliding and rocking down a wire; the conjunction of socks and pipe-cleaners to make "snuggies" [makeshift jackets to restrain the pigeons]; daily stuffings of the birds into socks and weighting; punchboards and carbon paper; photoelectric cells and lightbeams focused across large plates of thin celluloid; the two large

> plywood training units, four boxes per unit . . . with oblique main panels and slowly moving chambers; a centrifuge; nitrous oxide and metrazol treatments; a brace of vicious and intractable, but clever crows—one pecked out the eyes of a prize pigeon; never enough plastic wood and Deco to make up for the unobtainable gears, shafts, wires, relays, small motors impounded by the authorities [for the war effort].[88]

One of the surprising discoveries of working with pigeons was the varieties of behavior they could perform if held by hand and reinforced after pecking. While thus held, they "closed a circuit by pecking a small strip of translucent plastic exposed through a hole about an inch in diameter, an early version of the standard pigeon key." By "successive approximation"—that is, reinforcing a pigeon's pecking on a step-by-step basis—Skinner quickly ("in a matter of minutes") shaped a range of behaviors in the bird, from playing a simple tune on a four-key piano to batting a Ping-Pong ball. Skinner recalled the day he shaped the behavior of a pigeon that was hand-held rather than in a box as one of "great illumination." Skinner would use pigeons rather than rats, and keys rather than levers, in subsequent experiments. Project Pigeon strengthened his belief in the future of operant conditioning for more complex, even human, endeavors.[89]

Greatly encouraged by these early successes and the suggestion of pigeons' missile-guiding potential, Skinner approached Minnesota's Dean John Tate, who was involved in defense research. Tate was impressed with Skinner's news and contacted Richard Chace Tolman, brother of the psychologist Edward, who was connected with the National Defense Research Committee (NDRC). Tolman, however, was not convinced that pigeons could guide missiles under combat conditions (with anti-aircraft fire), and in May 1941 the work on pigeon guidance was discontinued.

The attack on Pearl Harbor seven months later rejuvenated interest in the project, and the university provided modest funding for more experiments. Tate continued to support the project, writing to Tolman of his confidence that "a bird's vision and head movement constitute an instrument for guidance which is probably superior to anything which can be produced by the hand of man."[90] Tolman remained unconvinced and could not justify any NDRC funding. Skinner believed that Tolman's skepticism may have sprung from the lack of respect he had for his brother's behavioral psychology. But Skinner may also have seemed an academic amateur to Tolman. His experiments to date had surely not approached anything resembling realistic conditions.

By sheer chance the project came to the attention of the Minneapolis-based General Mills Company. Someone who had been trying to get the firm to support submarine research alerted the right people to Skinner's work with pigeons, and an official in charge of research was able to convince the chairman of the board to fund more research on bird guidance. General Mills offered Skinner $5,000 to develop the device to the point where government funding would be feasible. By January 1942 the university granted him a sabbatical, and he was free to devote all his time to pigeon research. Elliott noted that Skinner "took with him to his new laboratory a little group of graduate students selected from among our best. . . . [T]he group lived, ate, and slept the project, and about that time Skinner seemed to some of us to become what might . . . be called 'exalted.' . . . The students . . . together with their leader . . . felt that ordinary pedestrian work of the kind that we were doing in our war-ridden university just didn't count."[91]

In September Skinner became in effect a full-time employee of General Mills. He moved all research activity to the top floor of a flour mill in downtown Minneapolis. There, with the help of his most talented students—Norman Guttman, Keller Breland, and William Estes—work began in earnest to develop a pigeon's target-pecking skills. (Nothing, however, was being done at General Mills to simulate a vehicle resembling the missile that the conditioned pigeons would guide. It was presumed that a suitable vehicle must already have been developed.) By late 1942 the pigeon crew had developed a bird technology that was effective with a wide range of targets.[92] Skinner again pressed for NDRC funding.

During the spring of 1943 several NDRC officials visited General Mills to investigate. What they saw encouraged them to award a $25,000 contract to General Mills for the development of a homing device. The handling of the birds proceeded in the summer, when they began to be controlled by a hydraulic pickup, later replaced with a pneumatic device. Yet these advances were not done with the assurance that the pigeons would be used to guide a particular type of weapon. Problems arose when Skinner and his group had difficulty getting correct information about the vehicle that would stand in for the missile. There were also problems in getting pictures of the target to use in training the pigeons. The NDRC was demanding accuracy, but was not providing the technical information that would allow the researchers to be sure of fulfilling government expectations. Nonetheless, by January Skinner believed that "the mechanical problem of translating the behavior of the birds into a usable signal seemed very near solution."[93]

NDRC observers agreed that Project Pigeon had a plausible homing device, and recommended an additional $30,000 to perfect it; but their recommendation was rejected by headquarters in Washington, D.C., with no explanations. Skinner later learned indirectly that there was concern about "phase lag"—the time it would take to readjust the course of the missile's flight to keep it on target—and worked to cut that time in half.[94]

Two meetings between Skinner's group and NDRC personnel took place in Cambridge, Massachusetts, after which the NDRC concluded that it was not possible to determine from the data whether the vehicle would be stable—that is, true to the target. Skinner admitted that, given the measurements the NDRC used, this was true, but that the reliability of the device itself was not in question. Nevertheless, Skinner was at the mercy of their decision: "It was clear to us that most of the men present had only the vaguest notion of the proposed system, and we left the meeting with the feeling that [their] decision, favorable or unfavorable, would have little to do with the facts of the case."[95]

The NDRC's formal report stated that continuing to fund this project would mean seriously delaying others that it claimed offered "more immediate promise of combat application." Skinner was visibly distraught and later surmised that "the spectacle of a living pigeon carrying out its assignment, no matter how beautifully, simply reminded the committee of how utterly fantastic our proposal was."[96] But he was more optimistic than ever about the practical potential for behavioral science. Guttman recalled that, despite the NDRC's rejection, Skinner was "going all out" with experiments based on the key-peck operant.[97]

The government's rejection of Project Pigeon involved more than a disagreement over technical reliability or the veracity of data, however. Between World Wars I and II, the United States had become the world leader in modern technology. By the early 1940s the national government was putting unprecedented resources into weapons development, including the secretive Manhattan Project that would develop the atomic bomb. Since Skinner's pigeon guidance emerged at a time when the United States was fashioning what Dwight D. Eisenhower would later call the military-industrial complex, it seemed a backwater endeavor compared to other wartime research.

Moreover, flying bomb or cruise missiles had been around since World War I, at least as tested devices. Gyroscope guidance was the heart of their instrumentation, and automatic feedback controls had been patented back in 1916. NDRC officials were probably hesitant to graft pigeon guidance onto such proven systems. If the gyroscope worked relatively well as is, why risk pigeon guidance?

In short, although it was new, behavioral technology must have seemed strangely old-fashioned to the NDRC. Using animal devices in an age of atomic development was bizarre, even atavistic, and as it applied to weapons research was certainly outside the American technological-production mainstream. Thus it was not so much the NDRC's shortsightedness or lack of interest as the realization that other technologies and systems of production controlled the final evaluation of Project Pigeon.[98]

PROJECT PIGEON ONCE AGAIN revealed to Skinner the importance of a hands-on approach to his discoveries. For all his theoretical gifts and interest in epistemology and verbal behavior, he remained under the control of his inventing skills.

From his arrival at Harvard in 1928 to his departure from Minnesota in 1945, Skinner's life had been, on one level, the simple story of a young man finding a career, marrying, and raising a family. But those were heady years for him. At twenty-four he had been cynical, unattached, and much under the influence of his parents and his undergraduate friends. At forty-one he was married with two children and committed to his own science and research program. Along the way he had become one of America's best-known young psychologists. Being able to find out something about the world no one else had yet grasped put Skinner under the control of the very discoveries he had seemed to make, at least in part, by accident. After his discoveries the world would never be perceived the same way again to him.

By 1945 Skinner was both humbled and emboldened by his discoveries—humbled because he understood how fortunate he had been to find a new behavioral analysis and emboldened because he was beginning to see how this new science could be used. The future of operant conditioning lay in tantalizing but treacherous cultural waters. Here Skinner was in one sense as naive as he had been about psychology when he entered Harvard. But then, unlike now, he had not been armed with a science that seemed to have the promise to help shape a better future.

6

The Social Inventor Emerges

Practically every company in the country is planning to make refrigerators and radios when the war is over. The conversion to baby-cribs would be easy and the field is a hell of a lot less crowded.

—B. F. Skinner to Cuthbert and Janet Daniel,
March 15, 1945

When they married, Yvonne knew that Fred wanted children. She assumed that they would have them, even though she was not overly fond of babies. Certainly the prospect of being a mother did little to encourage her dream of a literary career, and when her first child was born in 1938 Yvonne was uneasy and unprepared. She was "scared to death of Julie" when they came home from the hospital. She even called the pediatrician when the baby cried to find out what she should do.[1] Skinner found her fears "rather amusing," but claimed that Yvonne "left everything to me. When [Julie was] born I had to go out and buy diapers and all that sort of stuff. We didn't have a thing ready for [her]." Indeed, her father would take on many of the traditionally maternal chores, putting the kids to bed after reading them stories and getting them off to school in the morning.[2] Yvonne loved her daughters dearly, but she had not been raised in a household where domestic chores were routinely done and hence had little inclination to do them when her own children arrived.

But she became more receptive as the baby became a child and began to talk and develop. When Yvonne was pregnant with their second daughter, Deborah, Fred suggested "that we simplify the care of the baby. All that was needed during the early months was a clean, comfortable, warm and safe place for the baby, and that was the point of the baby tender."[3]

For years he had cared for his experimental animals—cleaning boxes and cages, preparing food. He had sought to simplify the tending of rats and pigeons, and doing the same for his own children seemed eminently sensible. A special crib to simplify their care complemented his earlier attempts to invent such devices as the silent-release box and an automatic pellet dispenser, which facilitated the care of his animals as well as yielding better experimental results. The "baby tender," as he called it, was not designed to bring behavior under the control of reinforcers or to record cumulative rates of response, as the Skinner box did. Nonetheless, here was a designed environment, another enclosed space (this one with a glass front), whose felicitous effects on the child were readily apparent. It was Skinner's first truly social invention.

So in the summer of 1944, fresh from the disappointment of Project Pigeon, Skinner quietly began to work on the invention that would bring him into the public eye. In the basement of their large house on Folwell Street in St. Paul was a playroom that he converted into a study and a shop. A sheet of heavy plywood mounted on top of two sawhorses served as his makeshift desk at one end; at the other, on a workbench, was an assortment of small tools salvaged from Project Pigeon. There he built a thermostatically controlled, enclosed crib with a safety-glass front and a stretched-canvas floor. This would be Deborah's home within a home for the first two and a half years of her life—although at no time was the tender her exclusive environment, as has been written. She was often taken out of it.

It had bothered Skinner that Julie had been restrained by the traditional infant paraphernalia, not only diapers and nightgowns, but sleeping "on a thick mattress covered with a pad and sheet, and was zipped into a flannel blanket, her head protruding through a collar, her arms in flipper-like sleeves. It was impossible for her to turn over."[4] During Project Pigeon the plight of the restrained bird that must be free to peck the image of a target had been a persistent concern. The baby tender would restrain and protect the infant while providing remarkable freedom of movement. Skinner described his invention as "a crib-sized living space" in which his daughter wore only a diaper: "It had sound-absorbing walls and a large picture window. Air entered filters at the bottom and, after being warmed and moistened, moved upward through and around the edge of tightly stretched canvas which served as a mattress. A strip of sheeting ten yards long passed over the canvas, a clean section of which could be cranked into place in a few seconds." Deborah was soon enjoying untraditional infant freedom of movement, "pushing up, rolling over and crawling. She breathed warm moist filtered air and her skin was never waterlogged with sweat or urine."[5] Although loud noises were muted inside the tender, she could be heard from any room in the house. A curtain could be pulled to shut out light.

But there was more behind Skinner's decision to build the crib than the advantages of simplified care. At the end of World War II American families faced new challenges. Millions of returning servicemen married and started families. The significant drop in infant mortality, the postponement of childbearing until after the war, and the postwar national prosperity all fueled the baby boom. After the hardships of the depression years and the deprivations of wartime, Americans seized the opportunity to pursue the dream of prosperity and family happiness with unprecedented zeal.[6]

Yet, for all its promise, the postwar American marriage was unusually stressful. Couples and parents had sought the professional assistance of marriage counselors and child-raising experts in the 1930s, but these trends gathered greater momentum after the war.[7] The astonishing popularity of Benjamin Spock's *Baby and Child Care* (1946) and the uproar over Alfred Kinsey's *Sexual Behavior in the Human Male* (1948) were dramatic testimonials to the new strains of raising children as well as to marital difficulties.

Since at least the late nineteenth century, American mothers had been regarded as the chief rearers of young children, a task of immense social and moral importance. Urban and industrial growth had made parenting a nearly exclusively female activity, as males left villages and farms to work in city factories and offices. During and after World War II mothers found themselves in isolation, denied the support of extended families and sometimes even of husbands, who might be overseas or just away at work. Child-care manuals of the era made much of the new maternal anxiety.[8] Life-threatening epidemics of polio and outbreaks of meningitis, as well as published accounts that stressed the emotional and psychological adjustment of children, led mothers to fear for their children's health, safety, and happiness.[9] In many ways the anxiety that Yvonne felt about raising babies reflected very real biological and cultural realities of the America to which she belonged.

Skinner was aware of this new uneasiness, but unlike Spock—who wished to reassure mothers that "you know more than you think you do"—Skinner emphasized changing contingencies in the infant's environment rather than interaction between mother and child. His device was a designed environment for infant care that would reinforce mother-child relations. He was creating a new world for the American baby, a world that *freed* both mother and child by providing a more effectively *controlled* infant environment.

Skinner was also aware of the commercial potential for his invention. Writing to his friends Cuthbert and Janet Daniel, he acknowledged that his "interest in it is (1) commercial (I wouldn't mind getting something out of it) and (2) psychological (it would permit a lot of real research on the importance of the first two years of character formation)."[10] As early as January 1945 he had tried to get General Mills interested in developing the tender for the emerging peacetime market. The company was unsure about how mothers would react to the product. Placing babies in a totally enclosed space might arouse feelings of unnatural separation from the child. Also, there was the legal risk of accidents, such as infants who might be electrocuted or suffocated. After investigating the potential market as well as the marketing difficulties, a company representative concluded that "this whole thing—which involves babies at a very tender age—is a very ticklish subject for a lot of novices to play with. One underdone baby, one frozen youngster, or one smothered child or something of that sort, charged to General Mills, could be a pretty bad thing from a publicity standpoint."[11]

Skinner then tried another way to attract attention for his invention. In the spring he wrote a short article, "Baby Care Can Be Modernized," describing the baby tender's advantages, and sent it to the *Ladies' Home Journal.* An editor replied that the magazine was interested but had some questions about potential problems. What about crib odors? Could the baby's cries be heard in the insulated contraption? What about the emotional health of the baby—would the parents play with the child enough to keep it happy? Skinner replied that odors were minimal because temperature control dried soiling quickly. The insulation was only partially soundproof and easily permitted parents to hear the baby. And since the baby tender relieved parents from constantly ministering to the infant's needs, playtime was likely to be more enjoyable for them.[12]

The *Journal* purchased the article for $750.[13] Thousands of parents would read about the baby tender—and hundreds, perhaps thousands, would want one. The article appeared in the October 1945 issue, under the title "Baby in a Box." It enumerated the various labor-saving advantages of the tender: no beds to be made or changed, time saved not having to change the baby's clothing as frequently, and the need for only a weekly bath, as the "baby's eyes, ears, and nostrils remain fresh and clean." Furthermore, the tender eliminated diaper rash, reduced disturbing noises, improved posture and skin condition, and babies had fewer colds. It also facilitated a daily routine that allowed the mother to plan her day in advance, as scheduling of feedings and naps were more easily arranged. The baby would be active in the tender and could be enjoyed by the mother at any time. Unrestrained in a thermostatically controlled environment, tender babies were healthier and happier, and mothers were free to love their babies—who were, in fact, now more lovable. Skinner concluded that "it is common practice to advise the troubled mother to be patient and tender and to enjoy her baby. . . . We need to go one step further and treat the mother with affection also. Simplified child care will give mother love a chance."[14] Within a month after the article's appearance, seventy-five to one hundred baby tenders were being built by *Journal* readers.[15]

Although the article advertised the baby tender and described its advantages, it also contributed to confusion and misunderstanding about the invention. When the editors changed Skinner's title to "Baby in a Box," they gave the impression that the author was as much a crackpot inventor as a progressively minded scientist, especially to those readers who only glanced at the article. As Skinner explained thirty-three years later, "The word 'box' . . . led to endless confusion because I had used another box in the study of operant conditioning. Many of those who had not read the article assumed that I was experimenting on our daughter as if she were a rat or a pigeon."[16] Furthermore, a box is often associated with death because of the shape of a coffin—an association that was especially strong during the war, when thousands of young Americans came home in boxlike coffins.

One Californian who had read about the "baby box" in the *Los Angeles*

Times was aghast about "this professor who thinks he can rear his little child by depriving her of social life, sun and fresh air": "How would this professor and his wife like to be shut in a glass box in the house all day? . . . They say they will probably keep her caged in this box until she is three. . . . It is the most ridiculous, crazy invention ever heard of. Caging this baby up like an animal just to relieve the Mother of a little more work."[17]

Another result of the article and subsequent interest in the baby tender was the perception that little Deborah was being harmed by the time she spent in the tender. Rumors circulated that she had become insane and/or committed suicide. As late as 1989 Skinner related that a former neighbor had mentioned a conversation in which someone had referred to "that daughter who killed herself." A psychology professor had passed the story on to his students.[18] When she grew older, Deborah Skinner tried to end the rumors herself. "People expect you to be 'a little crazy,' " she said, then explained: "My father is a warm and loving man. He was not experimenting on me."[19]* Over and over again, Skinner maintained that the baby tender was "a marvelous world for a baby and full of affection because it is so free of annoyances and demands."[21] He was adamant that "there was nothing mechanical about the care we gave our child. . . . It is possible to build a better world for a baby and the baby tender was a step in that direction."[22] Unfortunately, Skinner's and his daughter's attempts at correction did little to stop the misconceptions.

Yet the article received far more positive letters than negative ones. "I liked your scientific approach to the age-old problems of caring for a baby," wrote one reader. "I am preparing to build an air-conditioned crib similar to the one you built for Debby. The illustrations of her healthy, happy smile convinced us that a system like the one she used must be all right."[23] Skinner wrote the *Journal* promotional editor that "the most unexpected but most gratifying response has come from many people who say they will have another child if they can get such an apparatus. That puts the whole matter smack in the middle of world politics since it seems to be going to have an influence upon the birth rate."[24] Other letters suggested special uses for the invention, such as the care of baby chimpanzees.[25] Some writers suggested names for the invention, such as

*Deborah mentioned that the only lasting effects of her baby tender years were unusually prehensile toes and the habit of sleeping with only a sheet and no blankets.[20]

the "Boxinette."[26] One particularly enthusiastic letter came from Yvonne's mother: "All your relatives and friends and the University and even your in-laws have reason to be proud of you. All to whom I have shown the article (I have practically stopped people on the street) think it is wonderful."[27]

Skinner was enthusiastic about the future of his invention. The story was taken up by AP and UP dispatches. Pictures of the crib were published by RKO Pathe News; follow-up stories were appearing as far away as in the *London Daily Mail*; and descriptions of the crib were being broadcast by radio programs.[28]

A Cleveland businessman, J. (James) Weston Judd, offered to develop the baby tender commercially.[29] Eager to capitalize on the interest his invention was stirring up and having had no success with any other major firm, Skinner was more than willing to take him up. He wanted to make money, and he wanted American mothers and babies to enjoy the benefits of the baby tender.

Skinner's buoyant expectations for his invention may have been heightened by recent advances in his own financial and professional status. In December 1944, the board of trustees of Indiana University had approved his appointment as professor of psychology and chairman of the psychology department. The appointment would be effective the following September 1 for eleven months, at a salary of $7,500.[30] The University of Minnesota made a counteroffer in an attempt to keep Skinner, but he spurned it.

Yvonne considered Bloomington "a come-down even from Minneapolis" and, as if seeking a new identity, changed her name to Eve upon leaving Minnesota. She had never liked her name anyway.[31] Once again their friends would be faculty members, as Bloomington suffered from a sharp town-and-gown division. Nor was she impressed by the natural landscape of wooded, rolling hills. Outside of scenic Brown County, with a beautiful park, swimming pool, and restaurant, the rural surroundings did not seem as lush, prosperous, or charming as other parts of the Midwest. This move called for not only her own personal adjustment but also that of her two young children.

Fred enjoyed Indiana. His university duties required little teaching, and acting as chairman for half a dozen psychology professors was manageable, if not particularly rewarding. He was free to continue his research interests, write his language book, dispense administrative chores, and promote his baby tender.

Initially Skinner was impressed by Judd, whom he found enthusiastic and imaginative. Judd's corporate affiliation was Display Associates, whose letterhead advertised expertise in design and store planning and window and interior backgrounds.[32] The first thing Judd proposed was a commercial name for the invention. Neither Fred nor Eve had been satisfied with "baby tender" but hadn't been able to come up with anything else except "Baby Nook" and "Kiddy Korner."[33] Judd suggested a clever new name—Heir Conditioner—and hoped that it would become a household word.[34] Inquiries about the tender showed that the public liked the crib for itself, whatever it was called. What Skinner needed was promotional and entrepreneurial energy and talent. "There is definitely a big opportunity here for someone with initiative and vision," he wrote Judd.[35]

SKINNER'S PLAN WAS TO assign his inventor's rights to Judd's company, help develop and test a commercial model, and assist in publicity. In return he wanted to use a number of the company-produced cribs for research and to enjoy "a reasonable share" of the profits.

Display Associates was prepared to produce fifty Heir Conditioners at a time.[36] Skinner had found a businessman small enough to work with on a personal basis, yet efficient enough to produce cribs in quantity—a small entrepreneur with big plans. He would retail the product at $100 a unit. At that price the market would swell. Skinner invited him to Bloomington for further discussion. They were now "Jimmy" and "Fred."[37]

Eve, however, was suspicious. She wrote her uncle, George Green, a Cleveland attorney, to check Judd out. Green's reply was not reassuring. Apparently there was almost no information about Judd or his past. He worked at a credit store in Cleveland, and he and his wife were newcomers.[38] But Judd was coming to Bloomington and could be appraised up close. During his two-day stay with the Skinners, he reassured Skinner that Display Associates could rapidly produce a well-made crib. He was a draftsman and could now, after seeing the Skinner crib, make the necessary drawings to use as guides to production. Skinner had transferred the responsibility for building the Heir Conditioner to someone he barely knew, who had no capital, and who had no visible marketing success. Nonetheless, his confidence was running high, and another event seemed to bode well for the enterprise.

Sally Hope, the wife of the head of Indiana's art department and a new mother, wanted to raise her daughter, Sarah Jane, in a tender. Skinner offered to lend them "our summer model," which he had built for Deborah when she needed no more than 80 degrees of crib temperature.[39] Although there were some problems adjusting the thermostat to the November weather, the Hope baby was thriving in the summer crib when Sally's father, Julian Bobbs, arrived from Indianapolis to see his new granddaughter. Bobbs was a major partner in the Bobbs-Merrill publishing company and, upon learning of Skinner's plan for the tenders, wanted in on the venture.

Here, as if dropped from the sky, was the solution to Judd's shortage of capital. Skinner advised Judd to "take all the help Mr. Bobbs can give you if it doesn't cut too deeply into expected profits." It appeared they had found a well-heeled capitalist who balanced the "pioneering spirit" and the "humane side" with the profit motive.[40]

By the end of November, Judd had received ninety-three letters from people all over the United States and Canada who wanted to find out how to build or buy an Heir Conditioner. He projected a grand pace of production, so that a million-dollar-a-year business seemed only a few years distant. Meanwhile, Skinner began to worry that Bobbs might make "some sort of a merger deal." Skinner wanted to retain as much control as possible, so he wrote to his father for advice and asked him if he would like to get in on the venture. That way he could better ensure keeping the business in the family.[41] But William Skinner declined the offer. He had never been impressed with his son's abilities as a businessman.

An attorney for a Cleveland law firm drafted a letter of the terms of incorporation for the Heir Conditioner Corporation. Bobbs would receive 50 percent of the company's common stock in exchange for $6,000 in cash; Skinner would get 12½ percent for contributing $500, plus a credit of $1,000 for assigning his rights as inventor to the company; and Judd would receive 25 percent for a credit of $3,000 in return for promotional work and expenses. The remaining 12½ percent of common stock would go to an unknown investor who would pay in $1,500 in cash. Skinner and Bobbs sent their checks to the Cleveland Trust Company, which had established an account for the new corporation.[42]

Once his own money was invested, Skinner, pressured by Judd, began to worry more about Bobbs's controlling interest. Bobbs objected to Judd's managerial control, especially since Judd had invested no money while Bobbs was virtually subsidizing the whole venture. Skinner was as insensitive to Bobbs's position as major investor as he was blind to any fault in Judd.

By late February there were ominous signs that he had badly misjudged his Cleveland collaborator. Judd was not responding to prospective customer correspondence. Hundreds of letters had gone unanswered, and correspondents wrote Skinner asking why. But that was only the beginning. When Judd's first order, for Sally Hope, finally arrived in Bloomington in late March—after numerable delays and excuses—Skinner was shocked. The workmanship was shoddy—the Heir Conditioner was a monstrosity. Sensitive temperature regulation was essential, and Judd had installed a thermostat that was not designed to carry the required voltage to heat the crib properly. The thermostat was poorly installed and had a warning buzzer for overheating that would no doubt make infants jump out of their skin. Worse, the lamps that heated the crib were fully exposed. Their glow would be "psychologically objectionable, especially to the timid parent who has been scared of cooking her baby by skeptical friends." Exposed leads to the lamps are also dangerous; 110 volts would be sent through the hand when the light bulb got screwed in. Skinner's list went on and on: "The window through which the baby enters the conditioner should have no sharp or hard edges. A generous gasket of extruded rubber is necessary. In grasping a baby it is common to toss it lightly in the air to change one's grip. This would bump the baby's head in the present model. The baby can also fall against the side ledge when playing around."[43]

Adding urgency to Skinner's fast-fading confidence in Judd was the prospect of another company producing the Heir Conditioner. Daniel E. Caldemeyer of Evansville, Indiana, was an ex-basketball star at Indiana University who had recently inherited his father's business, the National Furniture Manufacturing Company. In early March, Caldemeyer had expressed an interest in producing the cribs and Skinner had advised him to write to Judd. Skinner now turned to Caldemeyer as a likely successor. The close proximity of the Evansville business would allow Skinner to monitor production.

Still, he was not quite ready to abandon Judd—testimony to both his naïveté and his sense of loyalty. If Judd could not manufacture cribs, he could at least promote them. Skinner wanted him to stop producing new Heir Conditioners and confine his manufacturing activities to working out an acceptable model "in *general* terms—i.e., size, weight limits, types of thermostat, etc." This information, along with a potential customer list and an attractive promotional brochure, was to be presented as soon as possible to Caldemeyer.[44]

The situation in Cleveland, however, had deteriorated far more than Skinner suspected. Several days after he made his recommendations to Judd, someone from Display Associates called to say that Judd had disappeared. Skinner left immediately for Cleveland, and what he saw there left no doubt in his mind that Judd was not at all what he initially appeared to be. The shop he had equipped to produce fifty Heir Conditioners could barely produce five. Only two models were actually under construction. Judd was being pursued by angry creditors, and his business associates at Display had severed all financial dealings with him. Judd had gone to Chicago, where he took an order for another crib—knowing full well that he would never build it. Far from being a hard worker who maintained late hours, Judd had apparently often curled up and napped in the shop. He had told Skinner that he and his wife were raising their own child in an Heir Conditioner; Skinner later learned that was not the case. But the worst of it was that Skinner now felt personally responsible for the fiasco Judd had created. He was also disappointed at the thought that he might not ever make any profit on the invention.*

By May, a month after severing ties with Judd, Caldemeyer agreed to develop a pilot model of the Heir Conditioner and to determine a marketing strategy. As if underscoring his complete disappointment in Judd, Skinner suggested changing the name to "aircrib": "The name 'Heir Conditioner' does not seem to wear well. It is a pun, and like all puns it can grow stale. Moreover it is not effective in word-of-mouth advertising because it is easily misunderstood."[46]

During the next year Caldemeyer and Skinner worked on developing

*Skinner felt the embarrassing effects of Judd's dishonesty for years after he had disappeared. In July 1948 the post office inspector in Cleveland wrote that during March 1948, "Mrs. C. F. Adams, Ardmore, Oklahoma, purchased an 'Heir Conditioner' from Mr. J. W. Judd . . . in connection with which a payment of $250.00 was made. To date the equipment has not been received, neither has her money been refunded." Addressing Skinner, the inspector said, "[F]urnish me with complete details concerning your knowledge of Mr. Judd's business and what connection, if any, you had with it."[45]

an aircrib that would be immediately acceptable to the public. One problem was finding a suitable material for the crib floor. Canvas tended to slack when wet, which made rolling on more dry canvas difficult. Skinner discovered that "lumite mattress" was something that could be washed clean and wiped dry. Even better, there was barely any odor even after several days without cleaning. Less flammable and more porous than canvas, lumite also eliminated the need for ventilation slots on the sides of the mattress. The only drawback was that it was hard to stretch out on a frame.[47]

Another obstacle was procuring the basic material used in the crib's frame, plywood, which was in high demand for the housing industry. Skinner suggested using tempered masonite instead, but Caldemeyer did not find it as workable a material. They decided to produce an all-steel aircrib, but then found that the price of steel had escalated.[48]

According to Caldemeyer's calculations, their aircribs would have to sell for about $420 each, "and I do not believe a retailer would sell very many at this price." In addition, his investigations revealed that they would probably have trouble getting insurance for it unless they added certain safety features that would completely eliminate the danger of suffocation. Caldemeyer backed out of the agreement when these issues came to light. He suggested that perhaps someplace like General Motors or a refrigerator manufacturer might be interested.[49]

Skinner was again disappointed, but still had faith that the aircrib could be effectively marketed by a small company. Later he believed he had made a crucial error in his dealings with Caldemeyer: "I never explained to him what the important features of the baby tender were." Hence Caldemeyer envisioned the aircrib as a room and designed it in metal, giving it the appearance of a big refrigerator.[50]

But Skinner's propensity to "make do" with simple materials was out of step with postwar consumer tastes. Deprived of consumer goods during the depression and war years, and in possession of considerable savings, Americans were eager to buy new, costly items—from automobiles and refrigerators to tract houses. An unprecedented consumer demand was emerging, and advertisers were making fortunes selling new products at prices Americans would have formerly found prohibitive.[51] It is impossible to know whether the public would have bought aircribs in quantity, but it is certainly possible that General Electric or Westinghouse might have been able to sell them with a price tag of $400. When offered the opportunity to produce aircribs, though, neither company accepted.

During the late 1940s and early 1950s Skinner continued to promote the aircrib, maintaining that it offered not only physical but psychological benefits. He complained, however, that "I have found the American businessman much less enterprising than he is supposed to be." Although several hundred cribs were being used by satisfied parents who had made them themselves, no large company would risk the venture. But his own plans for manufacturing the crib had been stymied, in part, because it had not been properly endorsed. Seeking to rectify this deficiency, he asked a physician at Childrens Hospital in Boston to sponsor his membership in the American Public Health Association.[52]

Prevailing product preferences also contributed to the failure to market the aircrib successfully. Women especially tended to shy away from "unglamorous devices," preferring "the conventional crib with all its frills and ribbons." One couple, having raised four children in an aircrib, were mortified by their friends' references to it as a "Robot Nurse," an "air-filtered fishbowl," and an "electrified cocoon."[53] Many other couples found in it the practical, beneficial "tender" Skinner had envisioned.

His final involvement with commercial production of the aircrib began in 1952, when John M. Gray, a user of the crib, suggested manufacturing them.[54] Gray had been a student at Worcester Polytechnic Institute and during the war had worked at the radiation laboratories of the Massachusetts Institute of Technology. After a stint in the armed forces in 1944, he served as an engineer at the Airborne Instrument laboratory on Long Island. At one time he had also operated an appliance sales and service store in Worcester. Gray was a man who seemed to have all the requisite skills to produce and promote the aircrib successfully. In March 1957, after extended promotional work, the Aircrib Corporation went into formal operation.

Gray took a grass-roots approach to the marketing problem. Well before the company was established, he sent out a questionnaire to over one hundred people who had built their own cribs from Skinner's specifications and reported its salutary effects. Gray generalized from the satisfied comments to strengthen advertising. A promotional flyer produced in the early 1960s noted that "the mothers of over two hundred babies" had enumerated the aircrib's advantages. The flyer also assured prospective customers that the device was not really untraditional: "It is like any crib. The AIRCRIB does not change the babies' normal routine of rides in its carriage, visits to friends and relatives, or play." And in explaining "why the Mother likes the Aircrib," the mothers surveyed reported that the crib reduced routine work and gave them more time to enjoy the baby. The flyer included a photograph of a baby in Gray's Aircrib contentedly nursing off a bottle held by its feet, with a teddy bear sitting nearby.[55]

From 1957 until Gray's death ten years later, the Aircrib Corporation produced and sold perhaps one thousand cribs. Skinner believed that the major difficulty in expanding the market was not parental uncertainty about the aircrib but the problem of getting large manufacturers to mass-produce them. Nonetheless, the device had been a part of the public domain since the publication of the *Ladies' Home Journal* article in 1945.

THE QUESTION OF why the aircrib never became a household commodity, as popular with American mothers as conventional cribs, bassinets, and baby carriages, involves more than corporate legal trepidation and the relatively high price of the product (Gray's model sold for about $350). One problem was the absence of a truly balanced investigation into comparative advantages and disadvantages of the device. Skinner admitted as much in 1968 when he wrote an interested party that "I have never carried out any comparisons of children raised in or out of the . . . Aircribs. It would be a very difficult thing to do." He was so sold on its advantages to the child, moreover, that he felt an experimental demonstration was not needed.[56] But it is precisely an impartial comparison by a respected, independent research organization that might have boosted public and corporate confidence. As matters stood, the "baby in a box" legacy still reigned. Neither the comments of satisfied users nor continuing publicity had much effect on the demand for the product.[57] When John Gray died in 1967, "so did his company," Skinner lamented.[58] And years later, Eve would say that the aircrib had been "a great nuisance because . . . everybody knew about it every place we went . . . and I got so sick of explaining what it was and what it wasn't."[59]

The failure of the aircrib to capture a national market may have had more to do with other postwar opportunities than corporate or public resistance. Caring for an infant would have been easier with an aircrib, but a house in suburbia *itself* seemed to present an attractive solution to most of the advantages that the aircrib promised—healthy children in spacious, clean living arrangements, the luxury of privacy for both children and parents, less labor for the mother with new home technology. American babies were indeed growing up in boxes, but certainly not the kind Skinner wanted.[60]

The aircrib had given Skinner a great deal of pleasure as well as worry and disappointment. By inventing it he had combined his love of hands-on activity and tinkering with his love for his children. He also invented a lesser-known device when Deborah was old enough to be toilet-trained. He called it a musical toilet seat: "We had a child's toilet seat equipped with a music box. The music box could be wound and triggered with a spring-mounted arm held by a piece of blotting paper or paper towel. As soon as the paper is wetted the music starts. We did this primarily as a warning to the mother that the child is ready to be taken off the toilet but it turned out to produce very quick sphincter control."[61]

He also devised a musical box with a cord and a ring hanging above Deborah's crib, but within her reach. "By pulling on the ring, she could advance the movement one step at a time and with a succession of pulls play 'Three Blind Mice.' It was evidently a very successful experience because she never tired of it. I only wish I could have changed the tune!"[62]

He maintained that he never experimented with his children, but they clearly, especially Deborah, evoked his inventive talent. Later she would be partly responsible for her father's fashioning of a device that would be the precursor of his teaching machine.

Skinner's ingenuity had come to the attention of the American public, albeit in a controversial way. He liked the limelight and certainly hoped his crib would be commercially successful, as well as a help to infants and mothers. Moreover, in making and promoting the aircrib, he glimpsed the potential power of his invention to change child-rearing practices. The matter of how children were to be raised was explored in another social invention in the spring and summer of 1945. This time, however, the device was literary and made Skinner a full-fledged designer of culture.

INDIANA UNIVERSITY'S psychology department had a longer experimental tradition than perhaps any in the country, and Skinner was looked upon by some as "by far the most original, independent, innovative figure in psychology" at the time.[63] Although he did little teaching, he was actively involved in the study of verbal behavior and in various research projects. Together with Sam Campbell, who was interested in negative reinforcement, he designed "a foolproof device to shock rats" that became commercially available: "I never liked shocking rats and imagine I missed some opportunities to make important discoveries because I could not bring myself to do so."[64]

For a time Skinner's psychological laboratory, located on the fifth floor of Science Hall, was presided over by an "unchallenged monarch," Walter Pigeon—a one-pound bird named after Skinner's mentor, Walter Hunter, that Skinner had taught to play piano. The Indiana student newspaper reported that one day "while Dr. Skinner was at his desk telephoning, Walter flew in the open window and perched calmly on the desk. Dr. Skinner, the phone in one hand, grabbed Walter with the other," and thus Walter's "career was launched."[65] Skinner also made do at Indiana by trapping pigeons outside the building by laying a trail of grain to an overturned box propped open by a stick and attached to a string. This method of acquiring experimental subjects was soon abandoned, however, as the birds carried parasites, which in turn spread to the laboratory rats.[66] Project Pigeon had proven the superiority of the pigeon over the rat as a laboratory subject, and after he left Indiana for Harvard in 1948 Skinner never again designed rat experiments.

An important professional development during these brief Indiana years was the organization, with the assistance of Fred Keller and Keller's colleague at Columbia University, William N. "Nat" Schoenfeld, of the Conference on the Experimental Analysis of Behavior (CEAB). Keller had taken a leave from teaching during the war to do Morse code research with the U.S. Army Signal Corps. But by 1945 he was back at Columbia, where he and Schoenfeld offered a beginning psychology course with "little or no history, no learning and unlearning of the mistakes of the past—a direct presentation of a science of behavior. The result [was] electric. . . . These students get the point at once and become original skillful thinkers about behavior."[67] In 1950 Keller and Schoenfeld coauthored *Principles of Psychology,* a text on experimental materials selected from a wide variety of sources. The text was so successful that more radical behaviorists learned operant conditioning from it than from Skinner's *The Behavior of Organisms.*[68]

The need for CEAB originated from a sense of isolation. Few professional journals were willing to publish articles regularly by operant scientists. Furthermore, both Keller at Columbia and Skinner at Indiana had interested, serious students of behavior who, unfortunately, had "no chance to reinforce each other." With the establishment of CEAB, "we did begin to develop a cult," recalled Skinner.[69] And Schoenfeld reminisced that "we had a new identity as an in-group, as a 'movement.' "[70] The conference was a series of informal meetings in the late 1940s in which operant and nonoperant camps, faculty and students, discussed issues of interest.

Though the Conference on the Experimental Analysis of Behavior itself was short-lived, it was the catalyst that ended the professional isolation of operant scientists: It soon became the Society for the Experimental Analysis of Behavior (SEAB) and later Division 25 of the American Psychological Association—a large organization within a massive organization dedicated to developing Skinner's operant conditioning. CEAB also served as the ancestor of the Association of Behavioral Analysis (ABA), which also met to promote operant conditioning. Skinner had mixed reactions to the organizational expansion of the originally modest and intimate CEAB: "We should have preferred to remain informal."[71]

A professional journal that would eagerly publish operant work was still missing. In April 1958 the first issue of the *Journal of the Experimental Analysis of Behavior* appeared. Ten years later its success led to the establishment of the *Journal of Applied Behavioral Analysis,* which published applications of behavior modification.[72]

In 1946, William Skinner purchased a house for Fred and Eve on South College Avenue, not far from Bloomington's town center. (Indeed, William Skinner continued to subsidize his son long after Fred was married.) Eve was still isolated and unhappy. Social life consisted of endless cocktail parties with other faculty members. She sought diversion but was bored, although she took great pleasure and interest in her growing children. She often read a novel or two a day. A special friendship illustrated how Eve's style of living diverged from Fred's, given the opportunity. Eve became companion and confidante to William "Bill" Verplank, a psychologist in Skinner's department. Verplank was a former student of Skinner's mentor, Walter Hunter, at Brown University and had come to Indiana after serving in the navy. Eve remembered that Bill "would come over and Fred would go to bed early and Bill and I would stay up all night. When Fred was out of town, I'd go to the dances with him and he was a very intimate friend and treated the kids well, too."[73]

When the opportunity came in 1948 to leave Indiana for Harvard, she was elated and later summed up the differences: "Harvard doesn't need people, Bloomington did."[74] But for Skinner Indiana had marked a great career shift. Recalling those years, he wrote, "I had concluded *The Behavior of Organisms* by saying 'Let him extrapolate who will.' Indiana was the point of my extrapolation. . . . I was on my way to something new."[75] Skinner was now as much the social inventor as the laboratory operant scientist.

7

A Design for Living

Plato didn't know how to produce a Republic; Bacon didn't know how to [produce] a New Atlantis; Cabet didn't know how to produce Icaria. In Walden Two, I said, "I've got some ways of getting something. Why don't we use them to produce a world that is wonderful?"

—Interview with B. F. Skinner, March 9, 1990

One of the few friends Eve enjoyed while in Minneapolis was Gretchen Pillsbury, the wife of Alfred Pillsbury, a giant in the city's milling industry. Mrs. Pillsbury reigned as the leading socialite, "whose love of drama and the theatre was only limited by her poor eyesight." In 1945 Eve read plays to her and the Skinners were drawn into the Pillsbury circle. One insider in the group was Hilda Butler, also a friend of Mike Elliott (and his wife, Mathilde), Skinner's chairman at the university. At a dinner party near the end of the war, Skinner sat next to Mrs. Butler, whose son and son-in-law were serving in the South Pacific:

> I began to talk about what young people would do when the war was over. What a shame, I said, that they should abandon their crusading spirit and come back only to fall into the old lockstep of American life—getting a job, marrying, renting an apartment, making a down payment on a car, hav[ing] a child or two. Hilda asked me what they should do instead, and I said they should experiment; they should explore new ways of living as people had done in the communities of the nineteenth century.[1]

As a child he had read stories about the Shakers and knew that Joseph Smith had dictated the Book of Mormon near Susquehanna. While at

Hamilton College he once visited Oneida, where John Humphrey Noyes had founded one of America's most successful utopian communities. He had read *Walden,* Thoreau's classic, about living alone in a self-built cabin beside a pond outside Concord, Massachusetts while a Junior Fellow and considered himself "a devotee of Thoreau ever since. . . . When I first bought a car, I carried a copy of *Walden* in it to remind me of the simpler life."[2] At the time of the Pillsbury dinner, he had just read a history of nineteenth-century American communalism.[3] He told Mrs. Butler that modern-day youth would have a better opportunity to build successful communities than their predecessors had. She urged him to write a book describing the possibilities for a better future. When he replied that he had already found his niche, she emphasized that "they still have to find theirs."[4]

In early June 1945 Skinner began writing about a fictional community as a better world for dissatisfied Americans. He typed the first draft in seven weeks (the only book he did not first write by hand), read it page by page to Eve, and revised sparingly (also in striking exception to his other books, which were written over and over again). He wrote in "a white heat" that seemed almost miraculous: "It was the most wonderful accident, and very close, indeed, to automatic writing."[5] He called the novel "The Sun Is But a Morning Star," taken from the last page of Thoreau's *Walden.* The manuscript was published three years later as *Walden Two.*

He believed he had written *Walden Two* in the spirit of its predecessor. Indeed, there are many similarities of emphasis in the two books: "I think a good deal of the thinking is Thoreauvian, particularly the possibility of working out a way of life independent of political action. My attitude toward punishment and aversive techniques of control fits nicely with Thoreau's civil disobedience."[6] And, like Thoreau's *Walden,* his book encouraged people to live simply. "In *Walden Two* there are very few automobiles, food is not purchased in small packages, tins and bottles to be thrown away, consumption does not require much heavy industry, a few copies of papers suffice for many readers, and so on. The citizens are not taking much out of nature nor are they putting much waste back in."[7]

Skinner structured his novel around a "standard utopian strategy": "A group of people would visit a community and hear it described and defended by a member."[8] The member is Frazier, the founder of Walden Two, a former student of a professor Burris at a nearby urban university.

Two returning servicemen, one of whom has also taken a course from Burris, come to the latter's office and mention that Frazier has started a new community. Burris and company (the two servicemen, their girlfriends, and Castle, a conservative professor of philosophy) are invited by Frazier to visit Walden Two. The rest of the book describes the communal life of Walden Two—a small society of about one thousand in which every adult works an average of four hours a day, regardless of profession. Labor credits are assigned in inverse relation to the popularity of the task—that is, there are more credits for cleaning lavatories than for preparing food. Property is owned in common and babies are raised in a nursery in aircribs and belong to the group, not to the natural parents. The economy is simple and largely agricultural. Work is not determined by gender, and competition is discouraged. There is little government, no formal religion, and ample opportunity to indulge hobbies and appreciate the arts. Walden Two has arrived at this social equilibrium by utilizing behavioral science, most pointedly positive reinforcement. An environment is arranged so that everyone in the community has the opportunity to engage in work, hobbies, and arts that are in themselves reinforcing. No one is paid a wage for doing work. The narrative proceeds as a dialogue among Frazier, Burris, and Castle—especially between Frazier and Castle, as the latter objects to almost every cultural practice in the community.

But if Skinner used a common utopian strategy to structure his novel, he did not use any utopian novel or, for that matter, any actual utopian community as a model for *Walden Two.* Indeed, he did not look at his fictional community as utopian in the sense of existing somewhere else in place and time. It was meant for the here and now, and the practices of the community were not fixed but always subject to modification should they prove unsatisfactory. Frazier emphasizes that there is "a constantly experimental attitude toward everything," and that solutions to the problems of living would seem to "follow almost miraculously."[9] Here was a place of perpetual tinkering and social invention, of never-ceasing improvisation. Nonetheless, Walden Two was *designed* and not, like operant conditioning, more or less accidentally discovered. Having found the laws of operant conditioning, Skinner was ready to apply them to humanity itself. In the introduction to the 1976 edition of *Walden Two,* he explained: "Those who know the importance of contingencies of reinforcement know how people can be led to discover the things they do best and the things from which they will get the greatest satisfaction."[10]

The novel was a bold extrapolation of the results of operant conditioning with animals, an imaginative effort to create a better way of life for humans. It was far more ambitious in scope than Project Pigeon or the aircrib, even though it was fictional. The novel was different because "from the very beginning the application of an experimental analysis of behavior was different." Skinner's voice, Frazier, maintains that behavioral science is superior to all traditional ways of improving the human condition, whether they be the wisdom of common sense, religious sanctions, the social rewards of capitalistic competition, or the socialistic redistribution of wealth. Frazier exclaims: "The one fact that I would cry from every housetop is this: The Good Life is waiting for us—here and now! We have the necessary techniques, both material and psychological, to create a full and satisfying life for everyone."[11]

But what were these "necessary techniques" that could provide an American way of life where people did work they found satisfying; had time for leisure activities like art, music, literature, and games; and had both privacy and conviviality without governmental, economic, or religious controls? They were the means of shaping behavior by *controlling* the contingencies of reinforcement of which individual behavior was a function. Since this had been done successfully with small organisms, and since humans were organisms, albeit more complex ones, the techniques of operant conditioning could be used to control individual behavior to enhance community life generally. This power was awesome and threatening to some. When Frazier asks Professor Castle, his most hostile critic, what he would do if he found himself in possession of an effective science of behavior, Castle answers that he would rather retain human freedom and "dump your science of behavior in the ocean."[12]

Frazier, however, argues that if Castle refused to allow a science of behavior to control people, "you would only be leaving control in other hands." And those other hands had historically proved themselves tyrannical. Indeed, the control of behavior had recently been used with catastrophic effect by the Nazis, and more subtly but with great success in politics, religion, education, and marketing. The fact was simple but profoundly important: The power to control human behavior had been held by the wrong people and used for the wrong reasons—personal aggrandizement—and its cruel effects had rained down on millions of people.[13]

Walden Two cast Skinner into dangerous cultural waters. He was no longer simply tinkering with apparatus or controlling the behavior of animals or trying to sell a social invention to improve infant care. Here was the behaviorist as social inventor in full bloom, one who dared to deny the *reality* of individual freedom, even though his experimental community would use positive reinforcement to preserve and extend the sense or *feeling* of personal freedom. Americans had valued their freedom because they had seen it threatened. Would the *value* of freedom disappear in a Walden Two–like world, a world designed to eliminate the threat of force? Were behavioral scientists merely wolves in sheeps' clothing? Was not Frazier really the leader of a new class of scientific demigods? Some Americans were not willing to allow Skinner to move his science from the operant box to Walden Two. This leap forward in scientific application strained traditional ways of viewing the long-held American belief in personal freedom.

HOW DID SKINNER COME to write *Walden Two* when he did? The conversation with Hilda Butler was certainly a catalyst. And during the war he had had influential conversations about the feasibility of applying behavioral science to the social world with various colleagues at Minnesota: the positivist philosopher Herbert Feigl; Alburey Castell, another philosopher, whom Skinner fictionalized as the conservative Professor Castle; Joseph Beach, an English professor; J. W. Miller, another philosopher; and the novelist Robert Penn Warren, who was for a short time at Minnesota. He even talked with two local architects, List and Winton Close, about the design of community buildings.[14]

The novel was also partly written for Eve, as an imaginary solution to her unhappiness in Minneapolis. In the introduction to the new edition, Skinner emphasized her dissatisfaction as part of a wider frustration: "I had seen my wife and her friends struggling to save themselves from domesticity, wincing as they printed 'housewife' on those blanks asking for occupation."[15] He was proud that the novel had a feminist theme, even though some women hated the book:

> The curious thing is that I wrote it in an effort to solve some of their [women's] problems—to give them a genuine equality and to free them from the traditional slavery implied by the role of women in Western Culture. Little thanks did I get! I have always explained it this way: If you free women of the responsibility of cooking, cleaning and bearing and taking care of children, they can see no reason why they should be loved. What a horrible thought![16]

Skinner also said that *Walden Two* was his attempt to solve not only Eve's problems but his own problems with Eve as well.[17] Eve, however, implied that the book generated more marital disharmony than it eased: "We had tremendous arguments about *Walden Two,*" she told an interviewer for *Time* magazine in 1971. Skinner confirmed Eve's opinion: "It was a feminist book . . . had men doing women's work. . . . But Eve resented it. She still hates the book."[18] And when Skinner was asked in 1979 why he did not join a Virginia commune inspired by *Walden Two,* he answered, "I'd have to get a divorce right away. . . . My wife doesn't believe in community."[19]

Whatever the effect of *Walden Two* on their marriage, the novel was an imaginative expression of his hopes for behavioral science as social invention. He had never doubted its application to humans: "I am quite sure that when I did work with rats and pigeons I was always imagining parallel cases in human behavior."[20] Less than a year after *Walden Two* appeared, he wrote the Harvard philosopher Donald C. Williams: "Either the application of human behavior is doomed to failure or it raises some new and crucial problems regarding relations among men. I don't believe we can solve these problems by clinging to some half-analyzed notion of individual responsibility."[21] By the mid-1940s Skinner was ready to challenge traditional cultural practices with a broadly conceived behaviorally engineered alternative.

There was reason for casting his social hopes for behavioral science in a novel rather than an extended treatise or an essay, as he later did in *Science and Human Behavior* (1953), *Beyond Freedom and Dignity* (1971), and *About Behaviorism* (1974): "I am frequently reinforced . . . for a sort of exhibitionistic wit. *Walden Two* gave me free rein in this style. It is not so much boasting as public confession."[22] His character, Frazier, could also "say things that I myself was not ready to say."[23] And all his characters "gave me a chance to blow off steam."[24] The novel provided a form in which he could freely critique accepted values from a behavioral scientist's point of view. It also finally gave him the opportunity to complete a work of fiction—something he had failed to do during his Dark Year.

Writing *Walden Two* also afforded Skinner a chance to examine his own long-standing problem of conceit, a personal characteristic he believed had been held in check by the humbling influences of marriage and Minnesota. By creating Frazier, who compared himself favorably with God and yet insisted that behavioral engineering made one self-effacing, Skinner could freely examine his own personality. Social invention in *Walden Two* was aimed at reconstructing not only society but also himself. He admitted as much to a correspondent in 1948 who had asked about his motivation for writing the novel: "I suppose that both Burris and Frazier are parts of me. Writing *Walden Two* was a sort of self-therapy in which the Burris side struggled to accept the Frazier side."[25] And Skinner's chairman, Mike Elliott, agreed that the two characters "are only slightly differentiated projections of Fred's own mind—Fred, the spirit, and Fred, the scientist."[26]

Skinner's spiritual side developed grandiose proportions toward the end of the novel. He later remarked that "the scene on the hill when Frazier lies in the position of a crucified Jesus has disturbed many people, but I wanted to deal with the God-complex which a man in his position would almost certainly suffer from."[27] Frazier's deification was a risky enactment of his own problem with conceit. Someone suggested that the novel was a transformation of the Skinner box into what amounted to a community box, a *box populi,* but Skinner wondered whether his conceit had also made it a "God Box."[28] The novel may also have been an attempt to isolate and control his own God complex. Indeed, Skinner often, mockingly but half-seriously, saw himself as a sort of savior to humanity.

Autobiographical tension in *Walden Two* was not so much between Burris and Frazier as it was in Frazier/Skinner's ongoing struggle between conceit and humility. And in the end humility was meant to win. Skinner called Walden Two a designed culture, whose designer, Frazier, has "arranged his own demise. No heroes. No philosopher-kings. No saints. No Führer, Duce, [or] Caudillo."[29] Skinner explained that "I deliberately made Frazier not a powerful contemporary figure. Few members of *Walden Two* knew about him. I was anxious to make it clear that the community did not survive because of a strong leader."[30] But he might have added that the liberties he took with the character brought back an echo of his mother's warning about "what people will think"—he did dedicate the novel to his parents.[31]

In broad social terms the problem of conceit related to the problem of who would control a particular social environment. Frazier, like Skinner, has inherited conceit from past conditioning: "Frazier is the product of one environment and he designs another. He designs it as *his* environment has determined he should. There is no other possibility. He will control who controls." Frazier's God complex and ultimate self-effacement depict a decisive historical juncture. Skinner wrote, "It seems to me that Frazier represents the confluence of a Judaeo-Christian or Protestant Ethic culture and Western science reaching at last human behavior."[32] His autobiographical confession in *Walden Two* was tied in with his belief that behavioral science would transform the problem of control in Western civilization. Although the novel touched revealingly on Skinner's marital and egotistical problems, it was never intended solely as a self-centered, therapeutic novel.

In 1945, however, there was a more immediate concern. Skinner had considerable difficulty finding a publisher. Mike Elliott did not believe "the book will have any success and possibly no publisher can be found for it."[33] Indeed, four large firms refused the book. Macmillan finally accepted it only after the author agreed to write an introductory psychology textbook, which appeared as *Science and Human Behavior* (1953). The rejection letters emphasized fictional heavy-handedness; editors felt the book was too intellectual, which, in turn, meant there was not enough action to sustain the reader's interest. An intellectually inclined novel by a little-known author could not be expected to sell well, either.

Edward Bellamy's utopian novel, *Looking Backward* (1888), had been an exception, having sold well in the 1890s—a decade in which thousands of Americans disillusioned with the inequities and labor strife of industrial capitalism found socialism appealing. Likewise, the sales of *Walden Two* did not soar until the 1960s, when there was also growing discontent with the prevailing social and economic order. *Walden Two* was born a decade or so before it found a market, one that was determined not by matters of flawed artistry or lack of action but by readers interested in contemporary social problems—especially the question of designing a better American culture. As one reviewer remarked, "Dr. Skinner may be a few hundred years ahead of his time, but he's in the groove."[34]

Several early reviews were encouraging. The *New York Herald Tribune* observed that "spinning utopias is one of mankind's oldest parlor games, and Professor Skinner has some ingenious new psychological twists."[35] The *Chicago Sunday Tribune* noted that *Walden Two* is "not a magic mountain, but it is a sunlit hill with an extensive view in many directions."[36] The *New York Times* published two divergent reactions. One reviewer found the novel "a brisk and thoughtful foray in search of peace of mind, security, and a certain amount of balm for burnt-fingered moderns"; while the other complained that "life in Mr. Skinner's land of milk and honey would be just as intolerable as any of its renowned predecessors."[37] The *New Yorker* judged the book "an extremely interesting discourse on the possibility of social organization."[38] These generally favorable initial comments reflected postwar readiness for diversion and relaxation. Travel books sold well at the time and so did science fiction, and with *Walden Two* readers could enjoy an imaginary community as if going there themselves on a literary vacation.

But there were some negative reactions, and Skinner was naturally concerned. He went so far as to try to get his old friend John "Hutch" Hutchens, then a senior review editor for the *New York Times,* to find a sympathetic reviewer.[39] But his editor at Macmillan, Charles Anderson, was enthusiastic about the initial reaction, and assured Skinner that "the publicity will be beneficial even though the reviewers quarrel with the book and all the ideas expressed therein. The early attention indicates they think it is an important book and their attention to it is very gratifying . . . for it was undertaken in spite of skepticism in some quarters."[40]

Skinner's unease, however, did not prove groundless. By far the most influential—and the most damning—review was published in *Life,* a magazine with broad national circulation. The reviewer, John K. Jessup, a *Life* editorial writer, did not mince words: "Boards of Planners unobtrusively tell every big and little Skinnerite exactly what he or she must do. Once they are trained, the inhabitants of *Walden Two* have 'freedom.' But it is the freedom of those Pavlovian dogs which are free to foam at the mouth whenever the 'dinner' bell invites them to a nonforthcoming meal." Such conditioning led to insidious dehumanization, the reviewer maintained, by eliminating "the very possibility of random personal choice." Jessup also linked Skinner's other social inventions with the novel. "The menace of the mechanical baby tender is nothing compared to the menace of books like *Walden Two.*"[41] In one stroke Jessup had made Skinner a Pavlovian (something, of course, he had for years argued he was not), suggested he was a bureaucratic dictator (a telling charge in a world only recently freed of Hitler and Mussolini), and implied he was a misguided scientist who would use his science to provide a conditioned slavery that masqueraded as freedom (an implication Skinner was never to overcome). Jessup quickly followed his *Life* review with one for *Fortune* magazine that conveyed essentially the same message.[42]

Among Jessup's complaints was one that especially irritated Skinner. Jessup objected vehemently to the novel's "presumptuous title." *Walden Two* was "as much like Thoreau's original title as a Quonset hut is like a comfortable and properly proportioned Cape Cod house." He had the right to invent a utopia, Jessup wrote,

> but what should really be held against him is the egregious liberty he has taken with the title of Henry David Thoreau's original *Walden.* For the truth of the matter is that Thoreau's book is profoundly antiutopian; it does not belong in the long line of antiseptic literature that began with Plato's *Republic.* Far from trying to escape into a "brave new world," Thoreau, the cosmic bum, set out resolutely to make the best of what he could right around home. . . .
>
> Briefly, Thoreau was perhaps the greatest exponent of old Yankee virtues of "use it up" and "make it do." He made a philosophy of the Here-and-Now, not the Far Away. . . .
>
> Books like *Walden Two,* then, are a slur upon a name, a corruption of an impulse. All Thoreauists will properly resent them. . . .[43]

Skinner had appropriated and misshapen an American classic and in so doing had revealed a clear and present danger. Jessup noted in the *Fortune* review that: "If social scientists share Professor Skinner's values—and many of them do—they can change the nature of Western Civilization more drastically than nuclear physicists and biochemists combined."[44]

Interestingly, Jessup's *Life* review had been written as an editorial and was unsigned. Only years later, through the Harvard economist John Kenneth Galbraith, did Skinner learn the identity of the author.[45] But the aggressive attack prompted a defense by Skinner, who counterargued that "not all Thoreauvians will agree that the title *Walden Two* is 'presumptuous.' The real slur upon Thoreau is to regard him . . . as an odd creature who roamed the woods and fields and kept a journal of what he saw." He continued:

> It was no "cosmic bum" who elaborated the policy of civil disobedience but an aggressive social thinker who would not accept without scrutiny any relation to the state. Many of the ideas of *Walden Two* are certainly due to Thoreau. Like the author of *Walden,* I am concerned with possibilities of the Here and Now, not the Far Away. Walden Two is no farther away from the rest of the world than Walden was from Concord. Nor did Thoreau go to Walden "to get away from it all" as *Life* contends. He went there to test a theory—the theory that a man can satisfy his needs in a simplified world in which he can also save his own soul. *Walden Two* is, similarly, not an escape, but a frontal attack upon the possibilities for a better world.

Finally, Skinner believed, the reviewer was mistaken in implying that "men are not now controlled. . . . *Walden Two* merely suggests that we try and put some sort of order into this control and see what happens."[46]

Years later Skinner noted that many people "objected to my reference to *Walden* in my title, but I think there are similarities. Both books recommend trying to build a world within the existing governmental framework. Both emphasize reducing personal possessions . . . [and] both represent experimentation in a design in one's own life."[47] Clearly, he believed his novel was as much within the American tradition as Thoreau's, which also distrusted large institutions and made economy a virtue while encouraging personal innovation. And he further maintained that most reviewers of *Walden Two* had frankly missed the book's point. When he eventually learned the *Life* reviewer's identity, he was not surprised also to learn that Jessup "regarded himself as a liberal but he would have burned every copy of *Walden Two* if he had had the chance."[48]

THE EDITORIAL SQUABBLE revealed a basic problem with Skinner. Where did he fit in the humanist tradition? Was he an innovative liberal or a dangerous conservative? How did his scientific applications to a better living environment fit into the American tradition? While many critics agreed that Thoreau's design for experimental living was not appropriate for everyone, few regarded *Walden Two* as appropriate for anyone.

During the early 1950s the most telling attack came from one of Skinner's most articulate and influential critics, Joseph Wood Krutch. Krutch taught at Columbia University and was an established biographer and critic. His most widely acclaimed book, *The Modern Temper* (1929), had strongly criticized the growing power of science in twentieth-century life, especially governmental manipulation of scientific expertise for nationalistic objectives. Interestingly, Krutch had published a well-received biography of Thoreau in 1948, the same year *Walden Two* appeared. After 1950 Krutch spent considerable time in Arizona, wrote books on desert ecology, and became something of a modern Thoreau, at least in terms of his literary subjects. Skinner had met Krutch in 1951 when they participated in a symposium on the social sciences, which was published the next year in *American Scholar.*[49] The latter's extensive critique of *Walden Two* appeared in his *The Measure of Man* (1954), which received the National Book Award for nonfiction. Skinner's protracted fight with "humanist" intellectuals really began in earnest with Krutch's attack on a scientific conception of humanity in *The Measure of Man.*

The uses of technology and their often catastrophic results had bothered Krutch for decades, but by the early 1950s he was especially concerned about the growing attraction for the "Science of Man." "Perhaps Hamlet was nearer right than Pavlov," he speculated. " 'How like a god!' is actually more appropriate than 'how like a dog! How like a rat! How like a machine!' "[50]

Walden Two was the most extreme contemporary fictional statement of a man-as-machine social world, as well as the culmination of a scientific determinism that had appeared in different guises from Descartes to Marx to Freud to Pavlov. "An analysis of Professor Skinner's thought will reveal very clearly in what direction the Science of Man is moving," Krutch believed. Skinner was "one of the most able and esteemed leaders in his field, and the author of a fantasy called *Walden Two* which describes the contented life led by the inmates of an institution—though Professor Skinner might dislike this designation—to which they have voluntarily committed themselves and where they are conditioned to like being conditioned."[51]

Unlike the philosophic, religious, and socialist utopias of the past, which despite their unrealism retained a noble ideal, *Walden Two* was an "Ignoble Utopia." In contrast to such perfected social worlds as Plato's *Republic* and Thomas More's *Utopia,* Krutch wrote, *Walden Two* did not rely upon human reason to bring the good life. Once the original leader, Frazier, had instituted scientific conditioning, the required behaviors, including thought control, would be in place for all future generations. The control that behavioral science established in *Walden Two* was so complete that the older view of human nature as containing some invariant "natural" qualities was superfluous. Krutch, however, preferred tradition: "To say that human nature cannot be changed means that human nature is something in itself and there is at least the possibility that part of this something is valuable. If we say that it cannot be changed we are saying that it cannot be completely corrupted; that it cannot be transformed into something which we would not recognize as human at all."[52]

But in *Walden Two* human nature is plastic; it can be changed "as much and in whatever direction you wish." Indeed, human nature no longer really exists in that novel; in its place is something Krutch called "The Minimal Man." Because there is no appeal to traditional standards of human nature such as "God's revealed word" or "the inner light of conscience," or to "the voice of nature," there is no check whatsoever on the possible emergence of an evil society: "Since no human nature capable of revolting against anything is now presumed to exist [in *Walden Two*], then some other experimenter—conditioned perhaps as the son of the commandant of a Nazi labor camp—might decide to develop a race of men who found nothing more delightful than inflicting suffering, and to establish a colony to be called Walden Three."[53]

Krutch wrote his critique of *Walden Two* less than a decade after the Nazi labor camps had been liberated, at a time when the United States was fighting communism in Korea and had been immersed in a cold war turned hot. Although never without qualification, he insisted on associating *Walden Two* with Nazi Germany and Soviet Russia. The manipulation of populations with brainwashing techniques seemed very close to Skinnerian conditioning, and even though Krutch recognized that the good life in *Walden Two* was more benign than life in modern totalitarian nations, he objected to the original seizure of power—whether by Hitler, Stalin, or Frazier—despite the fact that citizenship in *Walden Two* was completely voluntary. In each case humans were "subject to . . . [the dictators'] manipulations." Democracy has "no meaning or function unless it is . . . within a realm of freedom where the sanctions of democracy can arise." He asked readers "whether totalitarianism on either the model of Soviet Russia or Walden Two is what we wish for or must inevitably accept."[54]

Skinner appreciated the attention Krutch had given *Walden Two,* if not the message. In 1955, on sabbatical from Harvard at Putney, Vermont, Skinner was considering the feasibility of a real Walden Two. He wrote his former chairman at Minnesota, Mike Elliott, that "the Krutch book was, of course, a stimulus, especially since it is attracting attention to Walden Two." And it had helped Skinner focus more sharply on the essential problem of the contemporary world:

> I really think the central question of the twentieth-century is this: can man plan his own future or are there reasons why he must not be allowed to? If Krutch came across a culture like Walden Two, say on a mesa in Arizona—where people chose their own line of work, where children were educated for the life they were to lead, where music and art flourished, where economic problems were solved by designing a modest life à la Thoreau, he would be shouting to the house-tops, "This is it!" But what bothers him about Walden Two is simply that somebody planned it that way! Let the accidents of history work out a pattern and it's fine. Let someone try it as an experimental plan and that's evil.[55]

But Skinner also had a more politically inclined interpretation of Krutch's standpoint. Shortly after the latter's book appeared, he noted that "the attacks upon 'Walden Two' such as the recent one by Joseph Wood Krutch in 'The Measure of Man' are part of a general anti-scientific movement closely akin to, if not part of, the anti-intellectualism now rampant."[56] Skinner came close to identifying his humanist critics with the anti-Communist hysteria of McCarthyism. Indeed, behavior modification was thought of as brainwashing, which was a tactic, so the polemic went, of a Communist conspiracy to take over the world. Cold war emotions aside, Skinner had been pro-science since the late 1920s, agreeing then with Bertrand Russell and others who had argued that whether science benefited or harmed humanity depended upon its usage. Krutch, on the other hand, felt science had already dehumanized. He wanted to reemphasize the older humanistic-democratic tradition. But the tensions and fears generated in the early 1950s with the confrontation between the free world and the Communist world made the debate of science versus humanism at times less academically than politically charged.

In the summer of 1955 Skinner, along with Krutch, Erich Fromm, Reinhold Niebuhr, and George Shuster, had been invited to contribute to another issue of the *American Scholar.* Skinner prepared what became "Freedom and the Control of Men," which attempted to clarify questions about behavioral science as social engineering.[57] Krutch was one of the most effective critics in stimulating Skinner to sharpen and elaborate his own "humanist" position. Both Krutch and the clinical psychologist Carl

Rogers (who publicly debated Skinner in 1956) helped focus him along the lines of what eventually became *Beyond Freedom and Dignity*.[58] In the late 1950s Skinner assumed more directly the role of a social critic.

BY 1965 SKINNER had a list of 250 individuals who had expressed serious interest in joining a real Walden Two. Scattered efforts to start a behaviorally engineered community began almost from the moment the book appeared. In 1948 Skinner reported that a group of young people in Minneapolis was planning to operate an experimental community later that year.[59] In 1949 he wrote a prospective Walden Twoer about a community around Yale University: "These young people want to try some sort of community living this summer. . . . They need a small farm within commuting distance from a more urban center where some of them can get jobs." The correspondent was from Boston, and Skinner believed a similar effort could be made in that area. He asked the correspondent if he might know someone who would support an experimental community by contributing property for the group.[60] A communally inclined benefactor never materialized, but Skinner kept looking and some years later thought he had found one.*

By the mid-1950s he was brimming with enthusiasm about the feasibility of a real Walden Two: "I am sure it can be done, and that it will be the most reinforcing experiment of the century."[62] He wrote Arthur Gladstone, a Yale graduate student, that "interest in Walden Two continues, but nothing is very actively under way." A Harvard law student, Matthew Israel, "has at times been completely dedicated to some kind of cultural design, and we are having lunch together shortly so that he can tell me about a cooperative movement not too far from Boston."[63] Several of Israel's friends were trying to live communally on a farm near South Lincoln, Massachusetts, and for years Israel tried to generate explicit plans for a Walden Two. Gladstone, too, had set up in New Haven a cooperative house based on the novel. Skinner was interested enough to visit, and reported, "I spent one night there and saw some of the

*In the early 1960s Skinner believed that Owen Aldis, who had made considerable money in the stock market, would "back a Walden Two venture" as a way to "provide quarters for scholars on sabbatical leave." Aldis paid a company to find out how many academics would be interested. In the end, however, Aldis wanted a scheme in which members were paid a wage and in turn were charged for community services—rent, food, and so on. Skinner rejected the quite "capitalistic" enterprise.[61]

problems. Preparation of meals was passed around to all the inhabitants, and that meant very largely what came to be called 'fast foods'! I believe we had meatballs and spaghetti the night I was there."[64]

Contact with Israel and Gladstone and their communal ventures, along with correspondent inquiries about joining a real communitarian experiment, encouraged Skinner to think through in more detail potential problems in establishing a Walden Two. In early 1955, on sabbatical leave from Harvard, he had the leisure to do just that. Indeed, he continued in a way to write *Walden Two,* except now in the form of copious notes on matters the novel had avoided or only touched upon.

The setting was conducive. He lived in a room in the Putney Inn in Putney, Vermont, the site of the first Oneida community (which later moved to New York State). His daughters were nearby. Julie was in her third year at a Putney prep school, while Deborah attended fifth grade in an elementary school for local faculty and boarded with a faculty family. Putney, however, was not for Eve, and she left for Europe and a tour of Egypt. Skinner saw his daughters occasionally, but for the most part lived a simple, solitary life: "I had a large room with a single sunny window. It was over the kitchen and warmed mainly by the kitchen stove. On cold evenings I sat around in my overcoat or simply got into bed. I had breakfast and lunch in the dining room with a few regular patrons or an occasional visitor, and in the evenings ate snacks in my room. I drank no alcohol . . . and before going to sleep I listened to Bruckner."[65]

He had needed a sabbatical away from Harvard. The past decade had been especially taxing, as his schedule had been packed with administrative, professional, and intellectual chores and endeavors. Putney gave him time again to be the social inventor, a designer of living, the experimental engineer of the wonderful world he had devised in *Walden Two.* Further, it allowed him to reshape and manage his own way of life. "If not Walden Two at least a reasonable Walden One," he wrote to himself on the day he arrived at Putney. "I am resolved to construct a mode of living which will keep me in top condition for (1) finding out what I have to say and saying it and (2) enjoying music, literature, etc." He reflected that "fatigue is a ridiculous sort of hangover from too much reinforcement. As soon as I rest up—as I did last night, going to bed at 8 o'clock—I begin to bubble over with things to say, with leads to follow. Why isn't this the optimal life?"[66] Putney, like Thoreau's Walden and Skinner's Walden

Two, provided productive escape—a transformed daily routine that allowed living a healthy, unhurried, creative life.

In this comfortable setting Skinner composed dozens of notes in which he attempted to address the problems of establishing a real Walden Two–like community. The note titles show the range of his concerns, and include: Statements of Principles; Responsibility; Problem of Productive Use of Time; What the Community Guarantees the Individual; Financial Sanctions; A Name (he rejected "Walden Two" as being "too tied up with B. F. S."); Receiving a New Child in the Community; The Problem of Cleanliness; and The Contented Cow (an analysis of the efficient and healthy handling of community cattle).[67]

Writing notes was not enough. Skinner needed to visualize the physical setting of living conditions, and once again he became the hands-on inventor: "I drew building plans and worked out a way of manufacturing thermally efficient, double-walled concrete blocks which could be locked together to form walls."[68] For years he would keep a map of the fictional Walden Two in his study. And shortly after finishing the novel in Minnesota, he had built a papier-mâché model of the buildings and grounds. Interested colleagues were invited to the Skinner home in St. Paul, where he read from his novel while the model for it sat on the living room rug.[69]

INTEREST IN A REAL Walden Two persisted during the late 1950s. Israel continued to talk periodically to Skinner and began a short-lived newsletter, *Walden Two Bulletin,* which made suggestions for legally incorporating a new community.[70] But no such community was established. Sales of the novel grew slowly, but remained modest: 250 copies in 1955; 479 in 1956; 512 in 1957; and 880 in 1958.[71] By the early 1960s hopes for better sales were spectacularly realized. Only 9,000 copies of *Walden Two* had been sold between 1948 and 1960, but 8,000 were sold in 1961 alone. By the mid-1960s annual sales approached 50,000 and, in the early 1970s, doubled to 100,000.[72] By the time Skinner died in 1990, total sales had approached 2,500,000. Clearly something important had happened to effect such a sales revolution.

Part of the book's success was related to Skinner's rising public reputation: his development of a teaching machine (see chapter 8), innumerable talks in colleges and universities across the nation, appearances on television, and, of course, the controversy of another book, *Beyond Freedom and*

Dignity. But discontent with American culture also made his novel seem related to youthful unrest. The rumblings of the "beat" generation of the 1950s gave way to a broader upheaval, a counterculture that emerged mainly among white, middle-class American youth—especially college students, who aggressively questioned the style of living and values of their parents and yearned for an alternative to the "establishment."[73]

Walden Two certainly described an alternative culture, and as youthful discontent and book sales rose, so too did Skinner's correspondence. One letter writer took exception to the novel being "a corruption of an ideal," meaning Thoreau's *Walden.* Frazier's technology of control was better than an empty, illusionary freedom, she wrote. *Walden Two* was "cool," even though some of her friends had said they would rather die than live there.[74] Another writer claimed that the book impressed her more than anything she had ever read: "I have never felt a desire to commitment before." She wanted to live immediately in a real Walden Two.[75] One person was upset with the "hypocrisy," "administrative red tape," and "meaningless goals" of her university. She identified with Burris's decision to abandon his university teaching position and live with Frazier in Walden Two.[76] A seventeen-year-old referred to the "idiocies" of the society of the older generation and vowed not to raise her children in contemporary America. It was not "too early to begin wondering, planning, and working for a future" in a real Walden Two.[77]

Skinner also received letters that casually lumped *Walden Two* into a well-meaning but faddish American eclecticism that entertained almost any and every nostrum, so long as it provided an alternative salve to the maladies of mainstream culture. But he had never looked upon his novel as a "handbook for hippies" or a cultural catchall that would automatically make everybody, regardless of creed or cause, happy.[78] Some hippies mistakenly thought because Skinner's *Walden Two* got rid of government, it condoned acts like drug use and irresponsible sex. He also would have been disappointed to have filled a real Walden Two with misinformed eccentrics such as one correspondent who had confused Skinner with Pavlov: "I have hope that Pavlovians, . . . hypnotists, and new knowledge on the power of thought, new philosophies like Scientology and Dianetics, whether or not one agrees with them, will unite in a concentrated effort to help people to love one another instead of to destroy one another."[79]

How much better to receive the following letter from a young, middle-class family man who asked about joining a real Walden Two and took pains to explain why he felt he could benefit an experimental community:

> I'm not a foulball or deadbeat. I've lived and worked, as a carpenter-cabinet-maker, in this village since the end of W. W. 2. I have two kids in high school. Along with the bank, I own my own home. I write a column for the local weekly newspaper. I believe I have the respect of the community. So I figure I've got something tangible to offer a beginning of a utopian society. . . . I'm ready to go. So is my wife, if I find such a community.[80]

Skinner had worried about the kind of people best suited to begin a Walden Two. A hard-working, young American, dissatisfied with competitive, wasteful work, but also a skilled laborer, was an ideal candidate for such a community. What was not needed were self-indulgent, directionless, antiestablishment dropouts who simply took advantage of the counterculture for its mood-producing qualities.

As late as 1963, there were still no viable Walden Twos. A decade later, however, several small communities struggled to survive along the lines of Skinner's science, focused on shaping individual behavior by applying positive reinforcement to build noncompetitive, cooperative behavior.[81] Two communities, each with under fifty inhabitants, survive today. Both "Twin Oaks" near Louisa, Virginia, and "Los Horcones" (The Pillars) in Sonora, Mexico, have managed to remain mostly self-sufficient by selling, respectively, handmade hammocks and homemade cheese. Although most of the original members have departed and it is an ongoing struggle to keep adults from rejoining the larger society, both communities have achieved some stability, in the sense that they have an economic base, a modest history of success, and a continuing commitment to the application of reinforcement principles to encourage a behaviorally engineered social life.[82]

Of the two, Los Horcones has kept more closely to what Skinner had in mind in *Walden Two,* and has been most admired by other radical behaviorists. Techniques of shaping a noncompetitive environment by using positive reinforcement are at the heart of Los Horcones's attempts to achieve both social control and a better way of life. The community has been especially successful in reshaping troubled and mentally handicapped children. The key to transforming the children has been a policy of using no "aversive stimulation."[83]

As experimental communities emerged, Skinner was asked again and again why he did not join one. He admitted to one correspondent that he "would be quite happy in Walden Two or any community reasonably similar to it," and believed that "I should have pursued a different line if I had lived in Walden Two all my life."[84] Although tempted from time to time to "make a break and set up a Walden Two," he believed his major contribution was "to promote the science of behavior which, I believe, will in the long run make such ventures successful."[85] Skinner's intellectual ambitions could better be fulfilled if he stayed at his academic perch and so, unlike his fictional character Burris, Skinner remained in a university setting. Besides, he had always meant Walden Two to be initiated and shaped by young people, and by the time Twin Oaks and Los Horcones were viable communities he was in his sixties. In addition, there remained the problem of Eve, who had no intention of joining him in any such place.

Yet, none of these reasons could fully account for his refusal to do what Thoreau had done—to "live deliberately" in the environment he had described. After years of invitation he did manage to visit Twin Oaks in the fall of 1976. He wrote Kathleen Kincade, a founding member, that "I like what I saw very much." But he was troubled by the lack of entertainment and leisure activities for Twin Oaks inhabitants.[86]

Carl Rogers thought Skinner too much of an individualist actually to live in any Walden Two, regardless of how expertly designed. When asked about this by a correspondent, Skinner denied Rogers's conclusion—and in so doing, probably gave his most candid reason for not joining. "Rogers is wrong," Skinner wrote in 1967, the year Twin Oaks was launched. "Though I am essentially now doing nothing I do not like to do in order to earn a living, I should have been pleased all my life to settle for a few hours of manual labor each day to discharge my responsibilities." He added, however, a crucial qualifier: "Of course, I do need stimulation," he explained, "and a community of 1,000 would not be likely to have in it many other people interested in my field. Correspondence is not an adequate substitute. I might have to move around a bit from time to time—I can't really say."[87] The prospect of living in Walden Two must have revived bittersweet memories of Susquehanna, a personal culture he enjoyed but also one of limited horizons and mediocrity.

Utopias are often partial attempts to regain a faded past—in this case, perhaps the enjoyments, without the liabilities, of early-twentieth-century small-town America. By the mid-1970s Skinner was certain that urban

American life would not survive, and believed that contemporary metropolises would become extinct—cultural dinosaurs to be studied by archeologists and anthropologists. In order for reinforcement theory to work, it had to be applied on a workable scale. The village was the place to begin and, by implication, end. He envisioned strings of villages gradually replacing unworkable cities.[88] But it was not an America of new Susquehannas for which Skinner yearned. That environment had been based on cultural practices that punished as much as they reinforced. He fervently believed "you cannot design a community where everyone has everything [but you can] make people productive and make people happy."[89] And he was much pleased when members of a short-lived Walden Two–type experiment in Canada, the Dandelion Community, wrote: "Thank you again for your dedication to a less punishing world."[90] Such letters lifted him in a way that moving to the communities he had inspired could not have done.

In fact, during the 1950s, although Skinner was very involved in the enterprise of designing Walden Twos, he had already turned his attention to a different but related area of social invention. In 1967 he wrote to a *Walden Two* enthusiast: "If I were to write the book today, I would spend much more time on the problem of incentive. In the long run I think the techniques of education will replace those of punitive governments and economic motivation."[91] Whereas Walden Two was an interesting experiment, a dream of producing a wonderful, small community life, Skinner's most ambitious attempt to apply positive reinforcement to society at large was to reform—indeed, revolutionize—American education through a "technology of teaching."[92]

8

Educational Engineering

> *A country which annually produces millions of refrigerators, dishwashers, automatic washing machines, automatic clothes driers, and automatic garbage disposers can certainly afford the equipment necessary to educate its citizens to high standards of competence in the most effective way.*
>
> —B. F. Skinner, "The Science of Learning and the Art of Teaching," 1954

As early as 1946 Skinner had been mentioned as a possible appointee to the Harvard psychology department. The problem as Boring saw it was: "Did he have something new which he could give us? Would he be so temperamental that he would be a nuisance to us?"[1] These questions aside, an invitation was delivered to Skinner for the fall term of 1947 as William James Lecturer. Skinner's ten lectures covered the topic "Verbal Behavior: A Psychological Analysis." Delivered between October 10 and December 12, they persuaded the department to offer him a permanent position.[2] Boring discounted "certain inflexibilities and certain immaturities . . . obvious in Fred [because] his reputation is spreading in the USA."[3]

Skinner would return in 1948 to a department that had changed greatly since he left in 1936. Faculty members with clinical or sociological interests, such as Henry Murray and Gordon Allport, had long been dissatisfied with what they viewed as the narrow experimental emphasis of Boring and S. S. "Smitty" Stevens. In 1947 they left the psychology department and established a department of social relations. Skinner told Boring that if he came to Harvard, he would come as more than a mere professor. He wanted a laboratory for the now flourishing science of the experimental analysis of behavior. Boring agreed to spend $4,000 of department funds to purchase operant equipment and another $1,000 a year to maintain the laboratory.[4] Harvard, indeed, would be far more than a teaching position for Skinner; he wanted to use the institution's prestige to promote operant science and related social inventions.

For Fred and Eve the chance to move back east came as "a relief to us both."[5] They had not been particularly happy in Minneapolis, and Eve had felt especially isolated in Bloomington. The move would reconnect her with old friends like Martha Smith and Lou Mulligan, both living in the Cambridge area. Initially, the Skinners lived for awhile with Ken and Lou Mulligan in Wakefield, Massachusetts, about ten miles north of Cambridge. It was the closest experience to communal living they would have, as the two families shared living quarters and domestic chores. But eventually the commute to Harvard proved too taxing for Fred. Besides, neither he nor Eve had wanted suburban living. After Bloomington, Eve "wanted to live in the center of everything."[6] And Fred was "in the mood for a clean sweep."[7] This sense of starting afresh in an intellectually stimulating environment was also, paradoxically, indicative of their need to become settled, a need strengthened by having two daughters who as yet had known no permanent residence.

In 1950 Fred had an architect draw plans for a one-story house on a double lot purchased in the Larchmont section of Cambridge, about two miles from Harvard. In November the family moved into their new residence on a quiet cul-de-sac where Julie and Deborah would grow to maturity and their parents would pass through middle and old age. After Fred's death in 1990 Eve would continue to live in the house, which more than any other came to be home. It was a modest three-bedroom ranch with a basement study, a workshop, and a backyard pool, added in 1960 after the sale of the Monhegan cottage. The basement study became Skinner's thinking haven—a place where he wrote the personal notes, papers, articles, and books that further established and enhanced his reputation as a major American social inventor, critic, and intellectual.

Skinner's long-standing interest in verbal behavior was stimulated by observing his daughters' verbal development. Both he and Eve took a keen interest in their verbal performance. Skinner remembered three-year-old Julie "hearing a two-year-old make a lot of speech-like noises to me to which I occasionally replied as if it were speech, [saying,] 'I don't want you to talk to my daddy.' " He speculated, "Was the child's unintelligibility the cause of this objection. . . . Should it not have been 'I don't like you *to talk nonsense* (to my daddy or anyone else)?' "[8] Eve took pains to capture accurately Julie's verbal repertoire—"not because it contains any 'bright sayings' . . . but because it is typical." In a "Record of Julie's Speech During a Twenty Minute Period," she recorded her daughter's speech on her third birthday:

> Julie and Fred are in Julie's room. I sat in the living room taking down her speech in shorthand. Julie was pretending her bedroom slippers were sick and was putting them behind a chair in the corner.
>
> "That's where the sick ones go. This is not a sick one, this is where they are not sick."
>
> (Fred: "Are they all sick?")
>
> "Yes, there are three, four. . . . Now this is sick too. This is not sick and this is sick, though. This is not sick and this (her tricycle) is a bigger sickie. . . .

That evening, May 1, 1941, "we also went through the 2 volume Century Dictionary and estimated her vocabulary by taking 30 pages and marking down all the words she knew. We did this twice and the estimates were within 100 words of each other. It is fair to say she has a vocabulary of about 2800 words at least."[9]

Skinner also recalled Deborah's verbal unconventionality. When quite small, she said she had two sore places on her leg. "One itches and one ouches," she explained.[10] And while teaching Deborah to read, he "noticed a few examples of mirror-reading (incidentally, she is left-handed). She occasionally read *and* as *the,* and I noted that one word is a pretty fair mirror image of the other."[11]

Skinner's efforts to help his daughters by applying behavioral techniques, however, were never made with anything approaching the scientific precision of his work with rats and pigeons. Julie once remarked that she wished her father "had used a few more [behavioral] principles on me when I was growing up. There were a few things that could have been done better. [He] really didn't consciously apply principles, I don't think, when we were growing up."[12] Skinner agreed, but on at least one occasion he did outline some general "instructions" that well illustrate his emphasis on positive reinforcement. It is easy to imagine the sibling arguments that may have given rise to these instructions to Eve:

> DON'T MAKE ANY MORE REQUESTS OR GIVE ANY MORE COMMANDS THAN ABSOLUTELY NECESSARY. MAKE A REQUEST OR

> GIVE A COMMAND *ONLY ONCE,* BUT MAKE SURE YOU ARE HEARD. IF NO RESULT, *TAKE IMMEDIATE NON-VERBAL ACTION.* TAKE AWAY OBJECT CAUSING TROUBLE OR MOVE CHILD TO SOME OTHER PLACE. IF TROUBLE STILL PERSISTS, SEND CHILD TO HER ROOM FOR SPECIFIED TIME AND NOTIFY HER WHEN TIME IS UP. AVOID CALLING CHILD BAD. DON'T SAY "PLEASE BE GOOD" OR "THAT'S A BAD GIRL." REWARD GOOD BEHAVIOR WITH GENEROUS PRAISE AND AFFECTIONATE GESTURES.[13]

These guidelines reveal Skinner's sensitivity to his daughters. He realized arguments would naturally arise between them, but his dislike of aversive punishment and his understanding of the determinants of behavior shaped his response to these situations.

Indeed, it was not only his daughters and their personal educational difficulties to which he was attuned. In a memo to members of the Harvard psychology department in 1955, he criticized the way the department taught its graduate students. He contended that aversive and negative patterns of control dominated the students' lives: "We do not teach; we merely create a situation in which the student must learn or be damned." The students, confronted with daunting requirements and given little assistance or encouragement, experienced stress that interfered with their scholarly achievement.[14]

Deborah Skinner was the immediate catalyst for what Skinner would soon refer to as a "revolution in American education." Six years younger than Julie, Deborah had learned to read more slowly and generally was not as adept academically as her sister. Debbie was not so much intellectually inadequate as a slow maturer, which caused her considerable frustration and therefore troubled her father. One day in 1953, he visited her fourth-grade math class at Shady Hill, a private school attended by many of the Harvard faculty's children. Sitting at the back of the class, he observed:

> Students were at their desks solving a problem written on the blackboard. The teacher walked up and down the aisles, looking at their work, pointing to a mistake here and there. A few students soon finished and were impatiently idle. Others, with growing frustration, strained. Eventually the papers were collected to be taken home, graded, and returned the next day.

Armed with his behavioral science, he had a revelation that there had to be a better way to teach:

> I suddenly realized that something had to be done. Possibly through no fault of her own, the teacher was violating two fundamental principles: The students were not being told at once whether their work was right or wrong (a corrected paper seen twenty-four hours later could not act as a reinforcer), and they were all moving at the same pace regardless of preparation or ability.

The problem was: "how could a teacher reinforce the behavior of each of twenty or thirty students at the right time and on material for which he or she was just then ready?"[15] Teachers really did not know how student behavior was shaped, nor did they realize the importance of their own classroom behavior. If only every potential teacher could "shape the behavior of a small organism, like a pigeon, just to see learning take place in visible form!"[16]

But the problem went beyond ignorance of operant conditioning. Just as animal psychologists had watched their subjects and had been unable to make accurate observations because they had to take their measurements of behavior by hand, teachers needed technical help. In behavioral science, mechanization had made a real difference. Behavior had been traced and recorded through the use of instruments. If students were to learn immediately whether their responses were correct or incorrect, at their level of preparedness, a mechanical device was needed. A few days after sitting in on Debbie's class, Skinner built a primitive teaching machine. Richard Herrnstein, who in the early 1950s was a Harvard graduate student, recalled arriving early for a lecture when he found Skinner sitting alone, cutting up manila folders. When he asked what he was doing, Skinner replied that he was making a "model teaching machine." Herrnstein also remembered Skinner's deft, incredibly quick hand movements.[17]

Confronted with a problem, Skinner once again used the hands-on approach. And, as with the aircrib, mechanical invention was not simply making clever gadgetry. It was also social invention—in this case, the rudiments of what would become a "technology of teaching," which used the teaching machine as a device to facilitate programmed instruction (a series of questions programmed for the machine) and, most important, applied positive reinforcement to individual students in the classroom. Since the early 1930s Skinner had realized the crucial effect of immediate reinforcement in shaping behavior. Now he saw how the same techniques could be used in the classroom.

His first teaching machines based on the manila-folder model were known as Slider Machines and were used mostly to teach arithmetic and spelling. Math problems, for example, were printed on cards that the student placed in the machine. Next the student composed a two-digit answer by moving two levers. The right answer caused a light to appear in a hole in the card. In a second model, the student moved slider machines marked 0 through 9, and a number on each slider appeared through a hole in the card. The right answer caused lights to appear in a corresponding row of holes when a lever was pressed. When the student pressed the lever, it locked into place, thus preventing movement of the sliders to find the correct answer.[18]

Later he made a device that allowed students to compose answers to questions on a tape that emerged from the machine. Deborah Skinner remembered her father's earlier teaching machines as being "very crude." Problems were at first printed on pleated tape and later on cardboard disks.[19] A lever was moved that covered the student's answer with a Plexiglas plate—which prevented altering the answer and also revealed the correct answer. If the student answered correctly (which was often, as the questions were sequential and designed to ensure student success), another lever was moved, exposing the next question. If the question was answered incorrectly, the lever could not be moved to expose the next question and the student would have to try again.[20] The disk machine could not "read" a right or wrong answer. All it could do was cover the student's answer under the Plexiglas as it showed the correct response.[21] Questions were prepared as "frames," beginning with the simplest problems of spelling or arithmetic and proceeding gradually in degree of difficulty, so that a student seldom erred. Each correct answer acted as a reinforcement to encourage the student to proceed to a new question or problem.

SKINNER WAS NOT THE FIRST AMERICAN to invent a teaching machine. Dozens of patents on assorted mechanical teaching devices are on record at the U.S. Patent Office dating from the early nineteenth century. There were mechanical aids to teach subjects as varied as reading, the sense of touch, walking, dancing, boxing, and even a device to show visually how well a soldier squeezed a rifle trigger.[22] In 1926 Sidney Pressey, a professor at Ohio State University, became the first psychologist to build one. Pressey's device was a box containing a revolving drum. As the drum rotated a question or problem appeared in a slot in the box. The student selected one of four buttons as a response (in effect, a multiple-choice test). If the response was correct, the drum moved on to the next question; if not, the student pressed the other buttons to find the right answer and then the drum would move. One version of the machine dispensed candy as a reward. The student always found the correct answer, but Pressey's machines did not allow the student to compose an answer. They never really caught on in the 1920s. Skinner acknowledged Pressey's place as the inventor of the first effective teaching machine—even though Pressey did not program his device according to the principles of operant science, which were yet to be discovered.[23]

Skinner's initial public statement on the advantages of teaching with machines came at a conference on current trends in psychology at the University of Pittsburgh in the spring of 1954, and was published that year as "The Science of Learning and the Art of Teaching." "The simple fact is that as a mere reinforcing mechanism the teacher is out of date," he noted. "This would be true even if a single teacher devoted all her time to a single child, but her inadequacy [most teachers in the 1950s were female] is multiplied many-fold when she must serve as a reinforcing device to many children at once." There was only one sensible solution: "She must have the help of mechanical devices."[24]

The teaching machine, however, was only the instrument, not the *end,* of Skinner's educational revolution. It was a device that could deliver programmed instruction. The automation of teaching was as natural and desirable as any other popular and much-welcomed American technological advance, Skinner explained in his paper: "There is no reason why the school room should be any less mechanized than . . . the kitchen. A country which annually produces millions of refrigerators, dishwashers, automatic washing machines, automatic clothes driers, and automatic garbage disposers can surely afford the equipment necessary to educate its citizens to high standards of competence in the most effective way."[25]

But how did Skinner program a machine so that it could provide immediate reinforcement and individualized pacing for each student? Several features of programming were derived from his knowledge of verbal behavior. He began his work on programmed instruction while still finishing *Verbal Behavior* and believed that it was "the mediating step between the laboratory and education."[26] Human culture consisted largely of verbal behavior passed on from generation to generation. Thus the supreme social invention was verbal behavior itself, and the school was the public place where such behavior was crucial.

Verbal behaviors with special functions were given specific names and were discussed at length in Skinner's subtle and often difficult book. There were, however, three relatively simple techniques that could be used to mold verbal responses in the classroom: priming, prompting, and vanishing. *Priming* showed the student what to do, what directions to follow. Traditionally, the teacher either showed or told the student what to do. This provided a model to follow; it "primed" behavior so that it could occur. "After being sufficiently reinforced," Skinner noted, the behavior the student had been introduced to "occurred without priming."[27] But often priming was not enough. *Prompting* would supply part of an answer—enough to encourage the eventual mastery of the entire problem. One might, for example, provide part of an addition: 2 + __ = 4; or part of a quotation: "All good______come to the aid of their country." As he explained, "a prompt *hastens* the recall of a response in the presence of a stimulus which will eventually exert full control."[28]

Priming and prompting were common enough teaching techniques, but not as programs for teaching machines. Skinner tried his most original programming innovation on Deborah when she came home with an assignment to memorize lines from Longfellow's *Evangeline.* He wrote out a passage, had her read it, and then sent her out of the room. When she returned she found several letters from words erased, and she was asked to read the passage again. Reading it correctly, she was again sent from the room while Skinner erased more letters: "After five or six erasures, there was nothing on the board, but she 'read' the passage without a mistake." A month later she was the only one in her class who could still remember the entire passage. *Vanishing* was prompting in reverse, and Skinner soon incorporated it as programmed instruction in specially designed teaching machines: "I made several pocket teaching machines in which a passage to be memorized could be covered with sheets of lightly frosted plastic or clear sheets with obscuring spots or lines. The passage became less and less legible as additional sheets were placed over it."[29] Vanishing was especially useful in programming the vocabulary of foreign languages.

The teaching machine attracted the interest of some bright young behaviorists, who, with Skinner, began in 1954 to develop programming. Among them were Lloyd Homme, a former student of Skinner at Indiana who eventually sold an inexpensive teaching machine with Teaching Machines, Inc., in Albuquerque, New Mexico; John Carroll, who had worked with Skinner at Minnesota; Susan Markle, who would develop an arithmetic program for IBM; Irving Saltzman, who was on leave from Indiana University and would eventually become chairman of its psychology department; Douglas Porter, a graduate student from the Harvard School of Education; and Matthew Israel, the Harvard psychology student who had been so interested in developing a real Walden Two. Later, James Holland from the University of Pittsburgh became, along with Skinner, responsible for leading what eventually became a Committee on Programmed Instruction, and together they collaborated on a "Self-Tutoring Introduction to a Science of Behavior," published in 1961 as *The Analysis of Behavior.*[30]

In 1956 these enthusiastic behaviorists received a $25,000 grant from the Ford Foundation Fund for the Advancement of Education and some additional money from the Office of Human Resources. They were given room in an old Harvard building known as Batchelder House, where they tackled the problems of creating an instructional program. "How much of a subject should it cover? How much in a single session? How much in each 'frame' (as we began to call each presentation)? If frames were to reappear for reviews in later parts of a program, how should they be distributed? How much could we assume students already knew?"[31] These were fundamental considerations for creating a technology of teaching.

By 1957 Harvard officials were sufficiently impressed by the project to allow Skinner and his colleagues to use teaching machines on student subjects in the basement of Sever Hall in the Harvard Yard. During the fall term an instructional program was designed for Natural Science 114—Skinner's general education course based upon his 1953 book, *Science and Human Behavior.* The course was about the control of human behavior and aimed to get "undergraduates thinking the matter over in their own interest." Programmed instruction in this course used positive reinforcement to encourage students to answer questions, thus shaping competency in the Skinnerian subject matter. Enrollment for the class jumped 70 percent in one year, an increase Skinner attributed to the use of teaching machines.[32]

The Sever Hall experiment was encouraging, but success elsewhere made Skinner even more enthusiastic about programmed instruction. In the spring of 1960 he observed a class of thirty-four eighth-grade students in Roanoke, Virginia, using teaching machines to do ninth-grade algebra. Skinner never forgot what he saw: "The students were at work on the machines when we came in, and when I commented on the fact that they paid no attention to us, [Allen] Calvin [who had arranged the visit] went up to the teacher's platform, jumped in the air, and came down with a loud bang. Not a student looked up."[33] Moreover, the class completed in half a year a full year's work in ninth-grade algebra. When tested, they scored better than average for that grade; when tested again a year later, they showed a greater retention rate than a class taught by the usual methods. Clearly the teaching machine and programmed instruction were a boon for improving the math skills of American students.

By the mid-1950s large American corporations were showing an interest in manufacturing and marketing Skinner's teaching machine. International Business Machines was the first. Harvard's dean of the Graduate School of Education, Francis Keppel, had been impressed with Skinner's device and approached a member of IBM's board of directors who was a Harvard alumnus. In September 1954 Skinner received a letter from IBM expressing interest, and by November an oral agreement had been reached: "the company would look into the possibilities of manufacturing a machine (suitable for spelling as well as arithmetic), and I would try to find a foundation to support the development of programs."[34]

From the beginning, however, things did not go well with IBM. The company did not fully appreciate the uniqueness of the teaching machine and directed that it be manufactured in its electric typewriter division, then its major product. When IBM finally built a slider version of the teaching machine for arithmetic and spelling in elementary grades, it resembled a typewriter.[35] IBM was also looking into other technologies that utilized audiovisual concepts, such as the PRIVA-TUTOR, which would provide the student with practice in at least ten learning areas. Moreover, the company was embarking on computer development and by 1958 had constructed a device that combined a typewriter with a digital computer (the IBM 650) to teach binary arithmetic.[36]

During the late 1950s and well into the 1960s, Skinner had little respect for computers as teaching machines. They were expensive and did nothing to utilize the techniques of behavioral science. He was

Affectionate brothers: Fred and Ebbie, six years and two years, ca. 1910.
Courtesy of the B. F. Skinner Foundation.

Skinner in the late 1930s, at about the time of his marriage.
Courtesy of Hamilton College.

Unfinished model ship made by Skinner during his Dark Year in Scranton. The quality of the crafting clearly illustrates his ability to use his hands, a talent that would eventually facilitate the development of the Skinner box, the baby tender, and the teaching machine. *Courtesy of the B. F. Skinner Foundation.*

Skinner's mentor at Hamilton College, Arthur Percy Saunders, attending his prizewinning peonies in 1935.
Photo by Richard Carver Wood, courtesy of Hamilton College.

Yvonne, ca. 1936, when she and Skinner married.
Courtesy of the B. F. Skinner Foundation.

The young marrieds Fred and Yvonne, ca. 1937, in their Minneapolis apartment.
Courtesy of the B. F. Skinner Foundation.

Affectionate sisters: Julie and Deborah Skinner in 1945, shortly after the family moved to Indiana.

Originally published in Ladies Home Journal, *October, 1945. Photo courtesy of the B. F. Skinner Foundation.*

Eve playing with Deborah in a baby tender, 1945.
Originally published in Ladies Home Journal, *October, 1945. Photo courtesy of the B. F. Skinner Foundation.*

Deborah in a baby tender, in a photo included in a *Time* article.
Time Magazine.

A contemporary electrified Skinner box (*upper*) and its ancestor, the problem box (*lower*), which Skinner designed and Ralph Gerbrands built for Skinner's behaviorist friend Fred S. Keller in 1935. The early box was not electrified.

Upper: Courtesy of Gerbrands Corp., Arlington, MA.

Lower: Courtesy of the B. F. Skinner Foundation.

Darwin and Skinner caricatured in an 1871 cartoon and an illustration by Isadore Seltzer in the early 1970s. Skinner was always flattered by comparisons between Darwin and himself, especially with the implication that both natural selection and operant conditioning show how the environment has shaped both animals and man as parts of a common history of organisms.
Courtesy of Isadore Seltzer.

Skinner on the cover of *Time,* September 20, 1971.
Copyright 1971 Time Inc. Reprinted by permission.

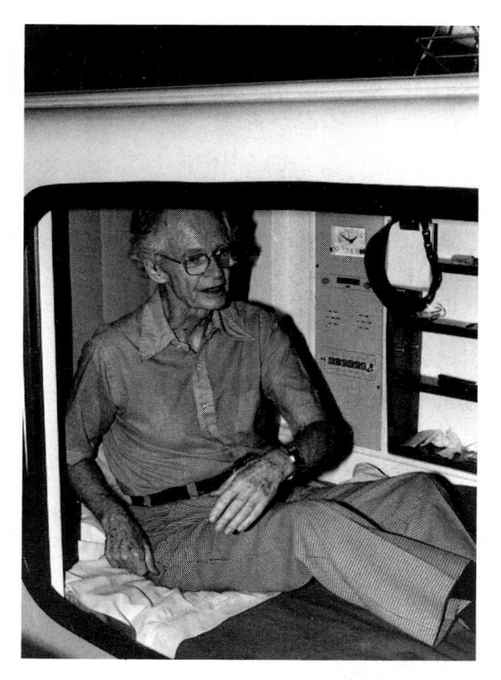

An elderly Skinner in the *beddoe* given to him by Japanese friends in the mid-1980s. The clock timer buzzed every morning at 5:00 to signal him to arise and move to his writing desk for the next two hours. The earphones next to the clock were used every afternoon to listen to his favorite composer, Richard Wagner.
Photograph by Dana Fineman/SYGMA.

Skinner at his writing desk in the late 1980s, opposite his closed cubicle. The large handwriting on the manuscript compensated for his poor eyesight. Skinner never learned to use a computer and used a typewriter only for writing *Walden Two.*
Photograph by Dana Fineman/SYGMA.

> also bothered by the attempt to humanize the computer. I once visited a company that demonstrated one such machine. I sat down and I pressed a button and the thing typed out . . . "Hello there. What is your name?" I wrote F-R-E-D, and it went traipsing back and typed "Hello Fred. We've been waiting for you. Now let's get to work." I knew perfectly well that that machine wasn't waiting for me, and I knew that when it said, "Hello Fred," it [was] no more real as a social response than the telephone operator's recorded voice which says "I'm sorry, but your call did not go through." No one was sorry my call had not gone through.[37]

The inventor and the company had two different technologies in mind; they were also working on different timetables. Skinner felt an urgency to market his teaching machine that IBM never shared. Moreover, IBM never properly acknowledged his role as the inventor. Skinner remarked to a friend who was trying to get Burroughs Corporation interested in producing the machine that "all [IBM] need[s] me for (and all they will pay me for) is the construction of material to be used in the machines. Nothing for the idea, nothing for testing models, nothing for prestige in the field—in short nothing for my good will. I had heard this of IBM but never quite understood."[38]

After testing ten model machines in a school in the spring of 1957, IBM put a spelling machine—the slider type—on sale in the fall of 1958. Yet no formal contract with the inventor had been signed. In fact, Skinner became so disenchanted that he approached another company, McGraw-Hill, that "was already marketing a device using cards with magnetic recordings on their backs to teach foreign languages." But McGraw-Hill also turned down the opportunity to produce the disk machine. By 1958 Skinner was working with Harcourt Brace, not only on the prospect of manufacturing teaching machines but also to publish programs. He was paid by Harcourt Brace as a consultant and that company began to produce a high school grammar program. In March 1959, however, the company's board of directors decided against producing teaching machines. Later that year Skinner terminated his relationship with IBM. By then he thought he had finally found a company capable of producing a more sophisticated teaching machine.

IN MAY THE RHEEM COMPANY, which manufactured steel drums, had approached Skinner. Interested in diversifying, they found the teaching machine an interesting possibility. Rheem's executive offices were in New York City, but Rheem Califone in Southern California was where the teaching devices would be manufactured. Skinner recalled meeting with a company representative "who painted a glowing picture of the financial rewards which lay ahead." In June a preliminary agreement was reached in which Rheem agreed to develop an improved model of the disk teaching machine Skinner was then using in his general education course, Natural Science 114. The machine Rheem manufactured, however, was again more like a typewriter, with students pressing keys to compose answers rather than writing answers on a slip of paper as they did in his course. When the formal agreement was signed in August, "the terms . . . were much less opulent than those we had first discussed."[39] It would be Skinner's last major commercial venture.

The strong refrain of disgust in Skinner's recollection of his relations with Rheem was similar to his eventual disappointment with J. Weston Judd and the Heir Conditioner some thirteen years earlier. But he had begun to sour on Rheem almost immediately. Skinner was considerably more skeptical about American companies than he had been in 1945, yet he needed American big business more than ever if the teaching machine was to alter American teaching practices significantly. The combination of disgust and need made for Skinner's most stressful experience as a social inventor. "I suffered from the treatment by businesses in the teaching machine era," he recalled, and would lay "awake nights writhing in anger."[40]

Finding the right name for his invention again became a problem. He remembered how the "baby in a box" motif had overpowered not only the overly cute "Heir Conditioner" but even the nicely descriptive "baby tender" and "aircrib." Calling it a "teaching machine" evoked the fear that it would replace teachers, even though Skinner emphasized again and again that it would only free them to help students in other ways. For the IBM machine he had used "Autostructer," which "unfortunately suggested a traffic accident." He remembered spending "many hours combining and recombining *self-* and *auto-* with all the roots I could find associated with teaching or learning."[41] Finally he decided on "Didak," derived from the Greek *didaktikos,* meaning "apt teacher," although he worried that it was too close to *didactic,*—a word denoting moral instruction or, worse, dictatorial control. One wonders whether *Didak* meant much of anything to those interested in the teaching machine. Certainly not what a Ford, Chrysler, or Buick meant to Americans interested in automobiles. Unlike the automobile, which was easily associated with various models, the teaching machine never found a popular trademark. Later, of course, the public would easily associate Apple with computer, but by then the computer, like the automobile, had succeeded as a generic invention.

Early models of Didak were exhibited at the American Psychological Association convention in Cincinnati in September 1958. As publicity grew, Skinner worried about his role in promoting the invention. He had been ambivalent about the negative association between the aircrib and the Skinner box, and as controversy grew about his science and inventions, he felt that it might be well to remove his name from his latest device. Seeking professional advice from his Harvard psychology colleagues, he shared a promotional announcement that read: "The Rheem Company announces the development of a series of Teaching Machines designed in consultation with B. F. Skinner."[42] Smitty Stevens merely replied, "I would go ahead as you have proposed."[43] But Eddie Newman responded more candidly: "If the machine turns out to be no good and it has been designed in consultation with you, you are the person to suffer. I think your strongest protection here is that the advertising must not overstate your contribution to the ultimate product." One took one's chances in dealing with large companies, a fact Skinner certainly appreciated.[44] It was agreed, however, that Rheem would use Skinner's name in the early stage of development and promotion of the Didak, but that Skinner's association would not be necessary once the product caught on.

As had been the case with Judd's Heir Conditioner and IBM's slider machine, Skinner was dissatisfied with the quality of Rheem's Didak, which had been hurriedly assembled and needed refinement.[45] Rheem executives lavished attention upon him, reassuring him that they would make the needed changes. In October 1959 the company's board of directors wined and dined him at a plush New York City private club. Shortly thereafter Skinner and Didak received national exposure on television when Skinner agreed to appear on Charles Collingwood's "Conquest Program." The opening sequence of the "What Makes Us Human" episode, which aired in 1960, showed students operating the Didak and, of course, continued the identification of Skinner with the teaching machine. But correspondence between Skinner and company officers revealed sharply diverging views of the Didak. It also showed the limits of Skinner's ability to control the quality and marketing of his invention, as well as his failure to appreciate the business point of view.

By April 1961 Skinner's frustration with Rheem peaked. The company had still failed to produce a device without defects. Moreover, it had ignored his suggestions for improvements and had "sent out [machines] for public testing" over his protest. As a result, Rheem had seriously damaged its corporate credibility and possibly Skinner's reputation as well. He also complained bitterly of the company's hesitancy in marketing the Didak: "What seems to me to be lacking is extensive and aggressive action. . . . I have already lost two years of valuable time in embarking on this development, and I do not think the Rheem company should try to hold me to our contract unless they are willing to take immediate action on an appropriate scale."[46] This sounded very much like his disgust with the various ventures to manufacture a viable baby tender.

Rheem's executive vice president was ameliorative, reminding Skinner of the company's "long-range position of strength with leading educational psychologists, curriculum specialists and school administrators"—hardly, however, people Skinner respected—which contrasted with "the 'quick buck' techniques currently exhibited by several firms in the field." The company assured him that it was still very much committed to improving the Didak.[47] But Skinner was not convinced. He wrote Rheem officials: "It is now two years from the signing of the agreement and Rheem has not yet produced an acceptable machine of any kind. During that period several other companies have successfully entered the field. I do not feel that Rheem can ask me to consent to further inactivity, and it seems best to protect myself against any further delay in the following way." Since Rheem was neither producing nor selling machines on a volume basis, he issued an ultimatum: "If this default is not cured within 90 days our agreement will terminate." Underlying the ultimatum was personal frustration: "I can no longer overlook the fact that I have been forced to remain inactive during important early years of a world-wide movement for which I myself am largely responsible."[48] Rheem had inexcusably delayed the hoped-for revolution in American education.

Rheem officials, however, saw the matter differently and believed they had been unfairly accused. They informed Skinner of all the steps they had taken to develop the Didak and judged that they had invested $300,000 on the project—more than three times the original estimate.

They also reminded him of the company's option to withdraw from the project at any time.[49] Thus, in essence, Rheem maintained that Skinner could not "fire" them because they could "quit."

Beyond differences in assessing what the company had or had not accomplished, or whether it had lived up to the contractual agreement, was a disagreement about Rheem's vital interests. Rheem reminded Skinner that "a prudent business enterprise must not commit itself to programs which it deems unduly anticipatory of market development."[50] And there it was: Rheem would not *risk* producing for a mass market until it was certain one existed. The company would not base marketing policy on the inventor's expectations or his leadership of a "world-wide movement"—no matter how well documented or well meaning. Skinner *knew* a market existed in every classroom in every school throughout America. If properly manufactured and marketed, the teaching machine and programmed instruction ("software" in computer language) would realize that need.

Skinner was unable to appreciate the corporate point of view and admitted as much: "I am not interested in business, have never consider[ed] giving much time to it, let alone all my time, and am willing to settle for the satisfaction of having started it all"—that is, the teaching machine movement.[51] Nonetheless, he worried that he had actually baited Rheem officials with his accusations.[52] He regretted his earlier "frankness, my wilted-shirt attitude, my willingness to go on . . . a first-name basis. If I stood on my dignity, kept advice in reserve, acted the 'professor' according to a businessman's script, I would probably be consulted and listened to."[53] But, as he knew, American business operated on criteria he simply did not appreciate. For months the relationship with Rheem dragged on in desultory fashion; the final legal rupture came in June 1963.

Skinner's assumption that American business was too conservative implied that this was not normal operating policy. Yet American companies traditionally had any number of possible technologies in which to invest. There were literally thousands of registered patents that hopeful inventors wanted companies to develop commercially. But stockholders required company officers to be accountable and, hence, to expand capital investment judiciously—for the nation's business past was riddled with the corpses of companies that had been *too* venturesome. He failed to realize that in America, especially in the twentieth century, safe enterprise was more American than free enterprise.[54] Skinner, in effect, believed in one of the beguiling myths about American business and suffered for it.

BEYOND HIS FAILURE to appreciate normally conservative American business practice lay another problem. Skinner's social vision was not empathic to modern industrial capitalism. He had the technocrat's vision of social benefit rather than the capitalist's acquisitive one. *Walden Two* had made it clear that a behaviorally designed culture would not be a corporate one that employed thousands of workers producing huge quantities of consumer goods. But this position was not something that began with the novel. His stance against commercialism was rooted in a sustained struggle as a young man to detach himself from the philistine world of his parents, from the unreflective anti-intellectual boosterism of Susquehanna and Scranton. His effort to carve out a special scientific and intellectual world distinct from that of his parents had been remarkably successful. Paradoxically, however, his success as a social inventor had necessitated a return to the Kiwanis Club culture he loathed. He now needed substantial financial backing from the philistines if he was to implement his industrial revolution in education. This was more than a matter of bringing an inventor and a company into a mutually satisfying relationship. For as Skinner dealt with IBM and Rheem, he found himself in conflict with his own personal history; he had to depend on representatives of an American way of life he had spent years repelling. His effort to relate to that corporate culture was the most frustrating experience of his public life.

Skinner was not simply an ivory tower intellectual with an elitist outlook. The teaching machine coupled with programmed instruction based on behavioral principles was a universal technology, meant to serve a mass society. He cared deeply about helping humanity, but to offer the social benefits of his science he had to court the world of boosterism and commercialism. As a man who had had remarkable success controlling experimental animals, he found it impossible to guide, reinforce, and control American business in the interest of his social invention. The commercial world could bring Skinner up short and enrage him, not only because of its broken promises and inefficiencies but also because it controlled the *social* territory that behavioral science wanted to reshape.

His high expectations and frustrations with American business were compounded by difficulties with the educational elite. Skinner took an especially dim view of "educational Specialists," by which he meant "educational psychologists, administrators, self-appointed reformers and many others. I simply must not publicly express my low opinion of them, for they are already sufficiently disposed to reject any help from a science of behavior." He considered them "badly educated . . . shaped by cheap successes . . . [with] a grim faith in the status quo. . . . They think metaphorically, illogically, or not at all. They assimilate a new idea to serve part of the established set and forget it. They are smug, unambitious."[55] Yet, if his machines were to be widely endorsed, he needed support from the very educational experts he loathed.

In late 1959 he came into contact with the consummate American educational specialist. James Bryant Conant headed a project funded by the Carnegie Foundation to investigate and improve the nation's high schools. Like Skinner, he was a Harvard graduate, but Conant's rise had been unprecedented. In 1933, the year Skinner became a Harvard Junior Fellow, Conant, then the forty-year-old chair of Harvard's chemistry department, was appointed president of the university, a post he held until his retirement in 1953. During World War II Conant moved in high government circles as an adviser for the Manhattan Project, and later became a member of the General Advisory Committee of the Atomic Energy Commission. For a short period he served as U.S. Ambassador to the Federal Republic of Germany. In 1957 he returned from Germany to begin, under the auspices of the Carnegie Foundation, a study of the American high school, which was published two years later.[56] The book appeared shortly after Americans had been shocked and even panicked by the launching of the Russian satellite *Sputnik,* which was a clear warning that the nation had fallen dangerously behind the Soviet Union in education—especially in mathematics and science.[57]

Why Skinner ever believed Conant would be enthusiastic about supporting teaching machines is something of a mystery. Even a cursory glance at his book, *The American High School Today,* reveals that Conant found the solution to educational problems in the high school to be mainly a matter of readjusting the educational bureaucracy. His emphasis was on questions of how many and how much. And Skinner realized that Conant "was not asking what I thought was the important question: How can teachers teach better?" Further, Conant reserved his most incisive criticism for the small American high school. Teachers there, he thought, could not "handle advanced subjects effectively"—by which he meant mathematics, science, and foreign languages. It was simply too expensive to hire qualified teachers for such small numbers of students, so "the elimination of the small high school by district reorganization" would solve the problem.[58] If Conant's suggestions were followed, Skinner's beloved Susquehanna High School might be one of thousands of tiny educational enclaves redistricted. His educational recommendations were symptoms of problems in American mass education rather than solutions to the crises such education generated.

Skinner had never liked Conant. His repugnance began in the mid-1930s when the just-appointed Harvard president refused to support Skinner's old mentor, William J. Crozier, in his plans to develop a general physiological department.[59]* Conant had also been close to people at the very pinnacle of political power in America. His Manhattan Project position had put him in contact with Roosevelt and Truman. Eisenhower had appointed him ambassador. Skinner was sensitive about academics who were close to the centers of governmental power, and national political leaders had never taken his behavioral science very seriously. Nonetheless, in late 1959, after signing the contract with Rheem and filming Charles Collingwood's "Conquest" program, Skinner felt upbeat about prospects for his invention and was willing to court Conant as a powerful supporter of the use of teaching machines in American schools. He asked for a meeting with Conant, which took place in New York on New Year's Eve, 1959.[61]

Their meeting was not immediately dispiriting to Skinner. Conant suggested that teaching machines and programmed instruction be tested in the Harlem school district. Harlem was a stable school system currently experiencing difficulty in teaching its students to read. Using teaching machines in the Harlem schools could provide dramatic evidence of their effectiveness. Skinner enthusiastically concurred. In addition, he hoped that financial support for such a project would come from the Office of

*The historian of science Philip Pauly has noted that Conant was an organic chemist who "looked askance at the physiological studies of photosynthesis that Crozier's protégés [at Harvard] . . . were undertaking." Moreover, noted Pauly, "while Crozier complained about the biological ignorance of chemists, Conant disdained Crozier's . . . generalized irresponsibility about 'controlling enzymes.' " Conant's view of science was more "classical"—meaning that he looked at the scientist as more an observer than a controller. Hence, his incompatibility with Skinner had a basis in fundamentally different conceptions about what scientists did.[60]

Education.[62] Skinner was so optimistic about the plan that he notified Rheem officials of the proposal.[63]

But he had badly misunderstood Conant's support for the Harlem project. Conant seemed not only to renege on the project but never to have been fully convinced of its viability. He wrote Skinner that it was necessary to submit the proposal "to a reading specialist in another city where there is a large Negro population and get his reaction. Without [this] first step, I do not feel I am in a position to be of much assistance to you, and I am not sure, of course, that even on the basis of your prospectus you could sell reading experts in very large cities on the advisability of trying out what you have in mind."[64]

Skinner recognized that Conant's support of the Harlem project could legitimize the teaching machine movement. Yet Conant also represented what Skinner despised in American education; and Conant's failure to endorse a workable teaching technology must have made Skinner wonder why he had gone to him in the first place. Conant feared that if the Harlem experiment went awry, students could be damaged; he may have also feared damage to his own reputation.[65]

Seeking to salvage the Harlem scheme, Skinner proposed a way to ensure the well-being of participating students while allowing the project to proceed: "The only answer seems to be to get experts [behavioral scientists] to evaluate what will actually be done to maximize the probability that the results will be at least harmless, if not favorable." Further, he maintained that any change in a school system that was failing so miserably had to be a change for the better.[66] But these arguments made no headway. Obviously the two men not only had different educational procedures in mind; they relied on different "experts." Moreover, they operated on different educational levels and had different educational theories and different educational agendas—to say nothing of the social distance between a famous dignitary and an ambitious social inventor. Skinner reported to Rheem that "I don't think there is a ghost of a chance any damage could be done [to the students]," but Conant's hasty retreat squelched the company's hope for an unqualified endorsement from a nationally renowned educational expert to boost Didak sales.[67]

As it became clearer that the revolution he had hoped for in American education was not at hand, Skinner recognized that his book, *The Technology of Teaching* (1968), which explained that revolution, would not be understood by educators.[68] "I find myself writing for experimental analysts," he noted. "I like to think that they will see the significance of [the teaching machine and programmed instruction]. . . . I am convinced that education as such cannot be changed."[69] By the 1970s Skinner believed it would be decades before a technology of teaching would make a significant difference in American education.

He did, however, change his mind about computers. "When I go back and look at the machines I invented," he told an interviewer in 1984, "I can see that they were just efforts to do mechanically what can now be done much more smoothly with computers. Of course, computers can do a lot of other things too. But as for the whole notion of presenting material and evaluating an answer, computers can perform beautifully now." Skinner had always emphasized the *program*—how most efficiently to change the *behavior* by appropriate steps or frames and to eliminate wrong answers that punished the learner—rather than merely the mechanical device. This was why he originally spoke in a derogatory way about the computer, which everyone else was hailing as a miracle worker. He still insisted that "the main thing is straight programmed instruction and the design of well-tested programs to teach basic subject matter. And that can be done without using all of the marvelous possibilities of the computer." Nonetheless, modern computers "could teach what is now taught in American schools in half the time with half the effort."[70]

He still worried, however, about those who used computers as games or toys. And he objected to the term "Computer-Assisted Instruction." "People speak of washing machines, not of machine-assisted washing," he insisted.[71] Computers should facilitate learning, so students would "gobble up their assignments" like they gobbled up the computerized bodies in the Pacman games.[72] Students liked good programmed instruction because they got things right and were reinforced by so doing.

While Skinner's ideas did not ultimately thrive in American education, neither did they completely die. The seeds he sowed are still alive and, to a limited extent, have been integrated in some computer-learning software or CAI (Computer-Assisted Intelligence).[73] But the educational revolution he had hoped for has not materialized. At the end of his life Skinner was discouraged about the future of American education: "Theories of learning . . . are destroying this country. Schools get worse and worse and all of the talk about that special teacher . . . doing a wonderful job, is no hope for the future. . . . I think education is the greatest disappointment in my life."[74] Teachers increasingly found themselves controlled by classroom environments unconducive to learning. They were not equipped with adequate behavioral techniques. Administrators and teachers alike were overwhelmed with maintaining student discipline. Students did not enjoy learning. A technology of teaching that could eliminate these difficulties was "being kept out of our schools by false theories of learning [and] teacher unions who are Luddites [a reference to the eighteenth-century English farmers who destroyed farm machinery for fear of losing their livelihood] and are afraid that this [new technology] is going to deprive them of their jobs."[75]

The educational establishment, by resisting behavioral technology, had failed to provide American children with adequate education. As a result these same educationalists unwittingly contributed to many of the social problems that plague American culture at the end of the twentieth century: drug abuse and its attendant crime, family disintegration, and the general decline of American urban life.

The teaching machine and programmed instruction gave Skinner more sustained personal and professional travail than any of his other social inventions. "I was . . . sick of the whole thing," he wrote in 1968. "How many scores of lectures and hundreds of conversations have I emitted in the past 15 years on the subject of teaching? I was desperate to bring it all to an end."[76] Through it all, however, he had tried to engineer a style of living that mitigated the adversive circumstances his teaching machine efforts had generated. "It is clear that even more drastic action is needed if I am to save myself for the things I want to do before I retire," he decided in early 1960. Among the possibilities were to "take on absolutely *no* further speaking engagements on teaching machines, no matter how important they may seem for the future of the movement." Another was to "let the Harvard Center for Programmed Instruction die aborning or live without me." And one was simply to "let Rheem go to hell."[77] The next year he observed, "My mood is, if anything, one of fatigue—I am tired of some of the things I have been doing and physically less energetic." Yet "I do not *suffer.* I seek rest and avoid exhausting labors, but I am usually *enjoying* my life—not ecstatically but steadily and contentedly."[78]

How did he manage to enjoy himself, though fatigued, frustrated, and even on occasion depressed? "Behaviorally I am active and my environmental design seems to keep me that way. I always have something to do (I planned it that way)," he explained."[79] And there were planned diversions. He and Eve shared a love of drama. Shortly after moving to Cambridge they joined a play-reading group, which selected plays and readers, edited parts, and provided stage settings. After rehearsal came an actual play reading, followed by refreshments. For over thirty years they belonged to this play-reading group, which made selections from Greek dramas to Shakespeare to modern offerings.[80]

Skinner always enjoyed jokes and even delighted in self-mockery. A 1960 satire entitled "The Psycho City Saga: A Psychological Elizabethan Western" showed that his lively sense of fun never collapsed. To the question, "What is psychology to me?" he responded:

The chair I sit in, its name is Edgar Pierce
I got it with reinforcement, and the competition's fierce.
I reinforced the trustees, with a good contingency
Now I'm a rich professor, that's psychology to me.

. . . . . .

First there was Helmholtz, then Darwin, Freud and James.
And now there is yours truly, he ranks with the greatest names.
And with my great elixir, I will all your troubles heal.
So gather round me brother, and with your savior kneel.[81]

Lighthearted diversions aside, Skinner was almost overwhelmed with too much to do: teaching, work with pigeons, attending committee meetings on behavioral research and educational policy, answering ever-increasing correspondence, and, above all, leading the teaching machine movement as inventor, consultant, program writer, lecturer, and author.[82] "I am still, I believe, recovering," he wrote in 1963. "My intellectual exhaustion during the past 6 years must have been very great."[83] Indeed, he was amazed that he had not suffered a heart attack: "I didn't have any physical symptoms that I know of except that I was so empty of energy by five o'clock in the afternoon at the office. I should have been warned by it. It was the right condition for a coronary."[84] Adding to his woe was the growing alienation from his Harvard colleagues, Boring, Stevens, Newman, Jerome Bruner, and George Miller, the last two of whom, especially Bruner, aggressively attacked the teaching machine movement.[85] By 1961 he admitted that he was "no longer interested in the [psychology] department. It has resisted all my efforts to improve it and is actually scandalously weak. It will be dominated for some time by intellectual fakes and smart-alecks."[86]

But, ever the behavioral engineer, Skinner would not succumb to burnout. By the mid-1960s he had arranged a daily schedule that, with a few changes, he would follow until he stopped going to his office in William James Hall once he learned he had leukemia, in late 1989. Skinner's account of his daily routine is worth quoting at length because it illustrates a workable, if obsessive, self-management during his time of troubles with the teaching machine:

> I rise sometime between 6 and 6:30 often after having heard the radio news. My breakfast, a dish of corn flakes, is on the kitchen table. Coffee is made automatically by the stove timer. I breakfast alone. . . . The morning papers *(Boston Globe* and *N.Y. Times)* arrive, thrown against the wall or door of the kitchen where I breakfast. I read the *Globe,* often saving the *Times* till later.
>
> At seven or so I go down to my study, a walnut-panelled room in our basement. My work desk is a long Scandinavian-modern table, with a set of shelves I made myself holding the works of BFS, notebooks and outlines of the book I am working on, dictionaries, word-books, etc. On my left the big Webster's *International* on a stand, on my right an open-top file containing all current and future manuscript materials. As I sit down, I turn on a special desk light. This starts a clock, which totalizes my time at my desk. Every twelve hours recorded on it, I plot a point on a curve, the slope of which shows my overall productivity. To the right of my desk is an electric organ, on which a few minutes each day I play Bach Chorales etc.
>
> Later in the morning I go to my office. These days I leave just before 10 so that Debbie can ride with me to her summer school class. Later, in cool weather, I will be walking—about 1 3/4 miles. In my office I open and answer mail, see people if necessary. Get away as soon as possible, usually in time for lunch at home. Afternoons are not profitably spent, working in [the] garden, swimming in our pool. Summers we often have friends in for a swim and drinks from 5 to 7 or possibly 8. Then dinner. Light reading. Little or no work. In bed by 9:30 or 10:00. I usually wake up for an hour or so during the night. I have a clip-board, paper pad and pencil (with a small flashlight attached to the board) for making notes at night. I am not an insomniac. I enjoy that nightly hour and make good use of it. I sleep alone.[87]

Skinner usually worked under five hours a day on writing and office chores. Activities such as gardening and swimming helped to keep him fit. Music and a vodka and tonic or two helped him relax. Plotting curves of time spent writing at his desk gave him an accumulated sense of accomplishment. Skinner carefully *arranged* his routine and his personal environment to reinforcing effect.

By late 1963 Skinner had put in place a daily routine that minimized the time he spent at Harvard and gave the early morning hours over to thinking and writing. The lion's share of his day was managed to ensure that these early hours were both enjoyable and productive. The huge effort, personal fatigue, and limited returns on the teaching-machine venture turned Skinner away from social invention and active involvement with organized research and relations with business and toward a more contemplative, solitary (although by no means antisocial) way of life. Nonetheless, his most visible and controversial public exposure was yet to come.

9

Beyond the American Tradition

Almost all the negative criticism I have received seems to me to involve vast misunderstandings of my position.

B. F. Skinner, Basement Archives, 1971

When Americans picked up their September 20, 1971, issue of *Time* magazine, they encountered a grim-looking, white-haired Skinner flanked by four insets depicting his gadgetry and social inventions—a man in the middle of his controversial creations. To the left of his forehead a pigeon pecked a Ping-Pong ball. To the right a rat pressed a lever. On his left shoulder a finger pressed the key of a teaching machine to answer the multiple 7 × 5. On his right was a bucolic Walden Two–like scene. Above his head the caption read: "B. F. Skinner says: WE CAN'T AFFORD FREEDOM." Inside was a feature story—"Skinner's Utopia: Panacea or Path to Hell?"—that included a biographical sketch of his small-town origins, his failed attempt to be a writer, and his controversial career as a behaviorist social inventor. The article highlighted his latest and most provocative book, a best-seller boldly, even shockingly, entitled *Beyond Freedom and Dignity.*[1]

His inventions, both mechanical and fictional, had, of course, attracted national attention since "Baby in a Box" had appeared in the *Ladies' Home Journal* in 1945. He had gained more renown with *Walden Two* and its rise to best-seller status in the counterculture atmosphere of the late 1960s. The teaching machine with programmed learning had also increased public interest. Nonetheless, until 1971 Skinner had not arrived at that level of fame or notoriety that convinced the editors of one of America's most popular weekly news magazines that his image had earned a place on its cover. When the moment arrived, Skinner was sensitive about the publicity the cover story would inevitably generate, and worried over his old nemesis, conceit. He had been recently portrayed as an egomaniac in a *New York Times* article and worried that *Time* would do further damage. Now on the verge of his greatest public exposure, he was at times obsessed with his image, vacillating between fears of being interpreted as a scientific egotist and elation that he was now thinking "bigger thoughts" than ever.[2]

There was an eleventh-hour crisis about when the *Time* cover story would appear. It had originally been scheduled for September 6, but national and international events intruded. Labor's reaction to Nixon's economic policies bumped Skinner off that cover in favor of of George Meany. The following week, September 13, was reserved for Maine's Edmund Muskie, the front-running Democrat for president in the next year's election. That left September 20 for Skinner, but when Nikita Khrushchev died on September 17 it appeared that, while the feature story on Skinner would be included, Khrushchev would be on the cover. In the end, however, it was too late to change covers. The next week the worst prison riot in American history erupted at Attica, and he would have been bumped again. Skinner was eager for the publicity—not only for the personal limelight but for the opportunity to make his new book, as well as the science of behavior, clear to the American public.

After the *Time* story appeared, Skinner was, for a short period, the hottest item on national and big-city talk shows. During the next several weeks he appeared on Martin Agronsky's "Evening Edition," "Mike Wallace at Large," "The Today Show," "The Dick Cavett Show," William F. Buckley's "Firing Line," and a dozen lesser-known but popular television shows. Within a month, millions of Americans had read or heard about B. F. Skinner and *Beyond Freedom and Dignity.* He was "completely swamped" by mail, telephone calls, and visits, and he received two or three requests a day to give lectures.[3] His privacy was disturbed. Strangers often asked to shake his hand in restaurants. He had, as one writer noted, "acquired the celebrity of a movie or TV star."[4] No American had ever written such a book. If baby tenders had worried some American parents, *Walden Two* had infuriated certain critics, and teaching machines had frightened the educational establishment, *Beyond Freedom and Dignity* was the final outrage.

But what was all the fuss about? What had Skinner written to ignite a vitriolic reaction? Why had the book "already stirred nation-wide debate through the force and shock of its ideas"?[5] In nine chapters and slightly more than two hundred pages, Skinner had proposed that the survival of the human species and hence the future of humankind could be best engineered by a science of behavior that decried the concept of individual freedom as the solution to the great problems of the day: the threat of nuclear war, overpopulation, and ecological ruin.

Yet he had said as much in *Walden Two,* with some critical reaction but hardly a firestorm. And the same point was made in *Science and Human Behavior* (1953), the textbook he had written and used in his undergraduate Natural Science 114 course at Harvard, with barely a ripple of public reaction. Indeed, in 1970 he noted that in that book "I made what now seems to me a fantastic extension of analysis to human behavior. . . . It forced me to think through a hundred issues."[6]

The polemical style of the new book, beginning with its title, angered and unsettled many readers. It also appeared at a crucial moment in the nation's history, when Vietnam War protesters, African-Americans, feminists, and gay rights activists, as well as libertarians and humanists, were insisting that it was precisely *more* individual freedom and dignity that was needed to counter an insensitive government, a murderous military machine, an exploitative capitalism, and widespread racism, sexism, and homophobia.

But there was something else in Skinner's book that burrowed into the marrow of American belief, something that said that if the American people continued to embrace a creed whose bedrock was the pursuit of individual freedom and dignity, they would have an impoverished future if they had any future at all. *Beyond Freedom and Dignity* took level aim at the way many Americans saw themselves—past, present, and future—and argued that their views were not simply misplaced but socially dangerous. The book challenged what most Americans believed it meant to be American. Skinner was not simply arguing about the *meaning* of freedom; he seemed to dismiss it as a dangerous illusion. None of the revered figures in the nation's political and intellectual tradition had risked doing that. From Thomas Jefferson and Abraham Lincoln to Franklin D. Roosevelt, John F. Kennedy, and Martin Luther King, Jr., national leaders had treasured the tradition of freedom as an American birthright to be protected and extended in all arenas. This tradition was further articulated by national intellectuals such as Ralph Waldo Emerson, Henry David Thoreau, William James, and John Dewey. At the heart of freedom were the rights of the individual, which might be modified in certain social crises such as war or economic depression, but which were still the essential American creed. Skinner dared disagree. People had believed in a fiction called the freedom of the "autonomous man." Previous generations had done the best they knew and had used the literature of freedom and dignity to counter oppressive or aversive political conditions. But they had been handicapped by the lack of a scientific analysis of behavior to illustrate how the environment controlled human activity.

This in itself was enough to stir considerable controversy. A close reading of the book revealed, though, that Skinner was doing far more than attacking "autonomous man" and the future utility of individual freedom as the basis upon which to build a future for the survival of the human species. He was also proposing a radically different way of understanding human nature. Traditionally, what was thought to distinguish humans from animals was consciousness. Skinner, far from denying its existence, argued that "consciousness is a social product. It is not only not the special field of autonomous man, it is not in the range of solitary man."[7] But consciousness, self-knowing, thinking, and all cognitive activity had always been attributed to the introspective powers of the individual, to self-centered consciousness—the most carefully guarded kingdom of autonomous man. Nonetheless, Skinner argued uncompromisingly that one's own thoughts were actually the product of an evolutionary history and a verbal environment that had shaped speech and so-called cognitive activity. Human nature as self-consciousness was not located inside the organism, inside the mind, inside the brain, in a free will, or in God. Human nature was nothing more or less than behavior, verbal and otherwise, controlled by evolutionary history and the contingencies of the environment.

It was so simple and yet so revolutionary—to many readers, revoltingly so. Skinner wrote: "A self is a repertoire of behavior appropriate to a given set of contingencies." This was a radically different kind of human nature, one that took away the originating action of the inner person. Indeed, in this view, the inner person was the imaginary creation of traditional philosophy and psychology as well as of the literature of freedom and dignity. Autonomous man had never existed, and it was crucial to use behavioral science to introduce cultural practices that dispelled belief in this fictional personage and replaced it with the scientifically engineered human behaviors that would best ensure species survival. Indeed, survival was the only value that could be used to justify a design for a culture. An ideology of individualism must be laid aside if species survival was to be secured. Skinner concluded, "We have not yet seen what man can make of man."[8] This final sentence in the book was intended as the behavioral scientist's hope for a better human future, but it has most often been interpreted as a threat to eliminate individual freedom.

The fuss generated in part from ignorance. Few of Skinner's readers knew much, if anything, about operant science. All of them, however, knew about the traditional subjects Skinner discussed or seemed to dismiss: individual freedom, dignity, inner consciousness, free will, and human nature itself. He had translated traditional subjects into the behaviorist perspective. Again and again, he used strange terms such as "contingencies of reinforcement," "aversive conditioning," "negative and positive reinforcement," and "repertoire of behavior." He had taken the familiar and made it unfamiliar; he had inverted what they knew was good—individual freedom and dignity—and made "control," another word used excessively, into a virtue. Now the individual was not only not in control but really did not exist, except as a locus of controlled behaviors. And perhaps worst of all, if the reader dared to look at the personal implication of Skinner's argument, *Beyond Freedom and Dignity* annihilated the person one thought oneself to be. Skinner's conviction that behavioral science could design a better culture struck many readers as beside the point. It called into question the desirability of wanting to survive without such traditional supports as free will and dignity. Hence the book could only be understood by many Americans in a way Skinner had not intended. In this sense, he transcended his reading public, and they naturally reacted by aggressively defending their traditional opinions.

THE STORY OF HOW SKINNER came to write the book that put him at the center of a firestorm of controversy began in the 1950s with criticism of *Walden Two* and his reaction to writers such as Joseph Wood Krutch, his exchanges with the psychologist Carl Rogers, and his mid-

1950s article "Freedom and the Control of Men"—an early response to humanist critics. The book also had its genesis in two invitations in the late 1950s: one an offer to participate in a Ford Foundation series of conversations (known as the Fund for the Republic) with well-known intellectuals—especially political scientists; and the other an opportunity to give four talks on cultural design, the Mead-Swing Lectures, at Oberlin College.[9]

Thus, although the teaching machine and programmed learning dominated Skinner's energies from the mid-1950s to the mid-1960s, his interest in the topic of cultural design never flagged. Rising sales of *Walden Two* helped. So did escalating interest in building real experimental communities. As he worked on ideas for solving the design problems of an actual Walden Two and wrote more generally about the social and political aspects of culture, he felt an emerging conviction that "freedom," as an intellectual concept, was not the solution to cultural survival, let alone the most beneficial of cultural practices.

Skinner's concern with the dimensions of analysis continued; only now, rather than being focused on the problem of studying the reflex—as he had been in the early 1930s—he concentrated on the perspectives most profitably used in studying societies: "I have always been concerned with the *dimensions* of things," he observed in 1968. "My thesis was in this vein. The dimensions of 'culture' are crucial."[10] By *dimensions* he meant something tangible, almost physical, in the sense that it could be sized up, measured, given shape and proportion. Culture was not so much a concept as a place that could be designed and built to behavioral specifications. In order to design a good culture, or even to visualize its design, one first had to find the dimensions of culture—its physical setting, its practices or behaviors. Fascination with the dimensions of something, whether a reflex or a culture, was a topographical or spatial way of thinking—an almost tactile hand-eye-mind perspective that well complemented Skinner's inventor modus operandi. "I like visual supports," he once observed. "I can still figure out the formula for extracting a square or cube root by recalling the blocks that were used to demonstrate the reasons for the formula."[11] Skinner was an intellectual architect concerned now with the building blocks of culture, of how culture is constructed; and to be so concerned was to think about the dimension of things.

Dimensional thinking about culture did not transform Skinner into a social scientist, historian, or expert on current social problems: "I'm not publishing anything on violence . . . , the revolt of the young . . . , drugs, the war in Vietnam, pollution, the revolt of the consumer. . . . I am talking about the *real* problems, not temporary manifestations."[12] He needed no model, theory, or ideology through which to write about the "*real* problems." He had a behavioral science from which to extrapolate a unique perspective. Here again was the behaviorist as social inventor, but also the social inventor as a special kind of social critic. *Beyond Freedom and Dignity* was Skinner's most forceful and comprehensive statement of his conviction that operant science could improve cultural practices. It revealed both his hopes and his fears for the human future. It was his most daring departure from his experimental science, although paradoxically based on that science's real experimental successes. He explained, "I am talking wholly in terms of behavioral processes. Much of the time it is interpretation only, but even so there are plausible experimental parallels or analogies." He was also going beyond other traditional intellectual precedents, beyond "Absolute Ideals . . . beyond the Utility, the Ideas, the Will . . . beyond the Existences and Structures of contemporary theory." The book's focus provided "something, at long last, to build on."[13]

One fellow behaviorist has observed that "Skinner tended to make exaggerated claims and then wait and see whether anyone would force him to back down."[14] In the past he had been challenged by Boring and the Pavlovians, but had held his scientific ground. Social inventions like the aircrib and the teaching machine were criticized, but he had defended their advantages. With the publication of *Beyond Freedom and Dignity,* he again courted criticism by maintaining that only operant science could properly shape human culture and that the notion of individual freedom was not only outdated but dangerous to the survival of the human species.

Skinner's own record of his struggle to write the manuscript that became *Beyond Freedom and Dignity* is illustrative of the writer's creative process, or what Skinner often referred to as "discovering what I have to say." Upon completing the book, he remarked: "The point of discovering what one has to say seems more and more important to me. Possibly I am improving my techniques of discovery."[15] This was especially interesting in light of the violent reaction to the book. Was there something in the *way* the book was created, not only its substance but its style, that infuriated critics to the point of preventing some from ever understanding it? Are there clues about what prompted the author to write and the audience to react as each did?

When he completed his manuscript in February 1971, Skinner was aware of the tremendous effort it had cost him to write it. "In these last days of work on *Freedom and Dignity* I find myself frequently groaning aloud. I cannot stop but I am in some kind of physical pain."[16] And "for about two and a half years—from June 1968 to January of 1971—I avoided all engagements which might interfere with *B. F. & D.* I was talking only to myself," he observed.[17] By the late 1960s Skinner had essentially become an American intellectual cloistered in his basement study, focused intently on writing a book that most other American intellectuals would vehemently reject.

First he outlined: "I have just rearranged the material from half-a-dozen papers into an outline for a book," he wrote in October 1964. "Since this is 'ripe' material, I have not run into the problems recently met in 'mining' my notes. I have been able—and have particularly tried—to emphasize a simple statement for the book as a whole." *Beyond Freedom and Dignity* was the only book that he was able to "think through as a whole."[18] He had grasped the major themes: "Controversy, costly in all science, takes the form in a science of behavior of complaining of a neglect of admirable personal features, such as freedom, responsibility, and dignity. The issue is controllability—only indirectly the existence of an inner controller." Furthermore, he observed, "the complaint about man's impotence to *control his own destiny* is answered by placing man *correctly in the causal stream.*"[19] The "causal stream" was located in environmental contingencies, not, as mentalists contended, inside man and in concepts of free will, accountability, and individual worth.

By 1965 Skinner knew the book would "cover a bit of the development of the evolution of cultures."[20] But he still wondered what specific subjects to discuss. He eventually decided to make the book more comprehensive and less vulnerable to changing scholastic trends, or "histories of thinking in these fields." He resolved to stick to "the *enduring* variables governing human behavior."[21] He would not discuss the development of government or religion as evolution through class conflict, or other functional or structural theories, the subject matter of social scientists.

Skinner had a low opinion of social scientists, even one of the most renowned. Having read in the *New York Review of Books* about Talcott

Parson's "theory of action," he observed: "I have often wondered what he had to say." The review quoted Parsons as saying that

> action consists of the structures and processes by which human beings form meaningful intentions and, more or less successfully, implement them to concrete situations. The word "meaningful" implies the symbolic or cultural level of representation and reference. Intentions and implementation taken together imply a disposition of the action system—individual or collective—to modify its relation to its situation or environment in an intended direction.

"So now I know," Skinner sarcastically exclaimed. This was verbose nonsense: "What is happening when a human being forms a meaningful intention and . . . implements it in a concrete situation? This is what *I* want to know."[22] A behavioral analysis of culture was not dependent on abstruse conceptualizations.

Social science had a teleological emphasis, being based on intentions and purposes rather than on the simple view of culture as an effect of determining environmental influences. As with his experiments with rats in the 1930s, Skinner tried to remove the superfluous complexities that made the study of behavior less reliable, controllable, and predictable. He strove for a comprehensive simplicity, with details subsumed in a behavioral topography that emphasized why some cultural practices survived and some did not. But his rejection of an eclectic, touch-all-the-bases approach as well as of the usual developmental perspective of social scientists was an intellectual strategy that earned him the ire of many critics whose scholarship and reading favored just such social science.

By 1968 he had clarified what would make his book distinctive. "I myself know how *Freedom and Dignity* differs from all the other books ever written on the subject. . . . It differs in that it is a discussion of behavioral processes, not essences, philosophies, issues, attitudes, ideas, etc." He would, therefore, "emphasize behavioral processes, teach my reader to spot them, formulate them, look for them, demand them." The stress on behavioral processes rather than intellectual abstractions was typical of Skinner. He did the same thing in psychology, by avoiding reference to mental states or to the structure and function of the mind. What he strove for was to make "the book intelligible to the non-specialist."[23]

A trip to England during the first five months of 1969 helped. He gave forty lectures there to students, psychologists, social scientists, psychologists, chemists, physicists, philosophers, linguists, and laymen. The audiences' favorable reactions sent him back home with renewed inspiration: "It will be a much better book. Generally I am much surer of my position. I have never been more confident that I am right, that my position is important, and that it will prevail."[24]

Many intelligent, educated people, he saw, were "hemmed in by intellectual hedges over which they cannot peek. They drive along familiar roads unaware of the green fields all around them. And the hedges?—logical, cognitive ways of thinking." His book would trim the hedges, allowing readers to see the behavioral "green fields." "I really *believe* it is possible to do something *essentially* different and better."[25]

As he approached the final stages of writing, confidence moved to certainty. "I find myself avoiding authorities in *Freedom and Dignity,*" he wrote in May 1970. His book was not going to be filled with " 'As so-and-so has put it . . . ' Why bother?" His resolve not to cite other authorities exemplified the old Skinner conceit. Indeed, there are few citations to other works in any of his books. Justifying the lack of authoritative references, he admitted that: "I am not a historian. I do not remember what I read, and I keep only sketchy notes."[26]

But there was another reason for avoiding authorities, a reason that critics would later identify as scientific arrogance or even fraud. Skinner explained that "I do not feel I need to appeal to authority. In fact I am saying that there has heretofore scarcely been any authority—before the advent of an experimental analysis."[27] It may have been arrogance; but it was also true that no such book had ever been written.

Because of the book's novelty, he worried that the intelligent reader, the nonspecialist, would not understand the jargon. Indeed, some fellow behaviorists found it initially difficult to comprehend his language and concepts. On the other hand, he realized that lay people had assimilated the jargon of thinkers like Sigmund Freud, Claude Lévi-Strauss, and even Noam Chomsky. The difficulty of being understood was mostly a matter of perspective: "Possibly my problem is that I am farther from lay psychology. In particular I get away from the familiar locale of mind, and into the easily forgotten territory of personal history."[28] Skinner's genius offered *new* concepts that allowed people to see their world in an entirely different way. They would, indeed, notice his book, if not understand it.

The stimulus for altering the book's original title, *Freedom and Dignity,* came from his editor at Knopf, who "pointed out that by the time I get through there is not much left of the traditional concepts. Almost at once

I suggested *Beyond Freedom and Dignity*. . . . There seemed at first something a bit corny about the 'beyond.' As if I were sponging on *Beyond Good and Evil* or *Beyond the Pleasure Principle*. But I like it. It now seems just right."[29] And the more he thought about it, the righter it seemed. He believed that *beyond* restored "true human achievement" by attributing the evolution of culture and its perpetuation to the selection of those consequences that allowed a future. The future lay beyond freedom and dignity. The problem was, however, that instead of reading *Beyond* Freedom and Dignity, most readers read *In Place of* Freedom and Dignity.[30]

•

THERE WERE SCORES of critical reactions to the book, from scholarly journals and newspapers to talk-show hosts. Perhaps the most notable of the latter was William Buckley's "Firing Line." Buckley's guests for the October 17, 1971, airing were Skinner and the English physicist Donald McKay of Keel University. In announcing the program, *TV Guide* called Skinner's ideas "the taming of mankind through a system of dog obedience schools for all."[31] Such an introduction wrongly equated operant conditioning with discipline and even punishment. It focused public attention on controversy rather than content.

During the program, McKay attacked Skinner's determinism rather than the major contention of *Beyond Freedom and Dignity*—that is, that scientific shaping of cultural practices that deemphasize autonomous man would bring a better future for humanity. McKay dominated the discussion, asserting that "freedom of action is a fact you can demonstrate if you think it through," which he proceeded to illustrate by digressions on the metaphysical proofs for freedom and a creative God. Skinner had little opportunity to restate the thesis of his book, let alone effectively to answer his detractor. He did manage to interject into McKay's monologue the argument that "the real fallacy is believing in . . . any of the . . . truth value systems which logic has proposed. We need an empirical logic which will be based on what men are actually doing. . . . This is a basic disagreement."[32] McKay quickly denied the fundamental nature of this insight and proceeded with his one-man show. But some viewers resented his overbearing verbosity. One irritated Skinner supporter wrote: "McKay's behavior backfired on him, and he did not come across as the Christian gentleman in practice that he seemed to be preaching about."[33] And Skinner felt that the "Scotch Presbyterian showed all the rudeness of a zealot and for some reason Buckley let him go on and on."[34] In tolerating McKay, Skinner had shown "that I suffer fools gladly."[35] The "Firing Line" episode epitomized a common critical approach to *Beyond Freedom and Dignity:* Critics had their own agendas and would not be content to discuss the book.

The most intellectually influential critique was Noam Chomsky's, for the *New York Review of Books,* on December 31, 1971. Chomsky, a linguist with an international reputation and a professor at Massachusetts Institute of Technology, had earlier written an exceedingly harsh review of *Verbal Behavior.* Skinner believed that Chomsky's demolition of that book had prevented it from even being read by linguists, thus crippling its impact. When he learned that Chomsky, whom the *New York Times* had called "arguably the most important intellectual alive," had reviewed his latest book, Skinner was not pleased.[36] As with *Verbal Behavior,* Chomsky would not review *Beyond Freedom and Dignity;* he would dismiss it and write about his own cognitive approach and political radicalism, an anarchistic suspicion of all governments, but especially those of the Soviet Union and the United States. In London for publicity engagements, Skinner noted: "Tonight will be a test. . . . I . . . learned that Chomsky has reviewed my book for the *New York Review,* as I knew he would. How shall I sleep tonight?"[37]

By the 1960s Chomsky had developed what he called "generative grammar," a complex analysis of how language structures emerged and developed from innate cognitive processes.[38] In contrast, Skinner's *Verbal Behavior* is not about the origin of the structure of language or even about language itself. Rather, it explains the *speaker's* verbal activity as an effect of environmental contingencies: audience response. Deep, innate, emerging grammatical structures versus environmentally determined speaking behaviors indicated more than fundamental intellectual differences. One man was a scholar with an activist politics, which above all valued human freedom; the other was a scientist who had just written a book devaluing political activism and the pursuit of individual freedom as the most effective way of ensuring humanity's future.

Skinner maintained in *Beyond Freedom and Dignity* that cultures had survived quite well throughout most of human history without much political freedom, thus freedom was not necessary for the survival of the species. The removal of political oppression had, nonetheless, been valuable and had promoted cultural practices that made social and individual progress more likely. Now, however, "autonomous man" and belief in "the literature of freedom and dignity" were blocking scientific solutions to problems that threatened the future of the world.* Behavioral science through operant conditioning had discovered techniques of positive reinforcement that could shape repertoires of individual behaviors that preserved and promoted humanity. Environmental contingences could be arranged to limit human population, to preserve environmental habitats, to avoid war. Individuals would not be "sacrificed" for these endeavors, but rather reinforced to behave in the interest of greater cultural benefits. "An experimental analysis shifts the determination of behavior from autonomous man to the environment—an environment responsible both for the evolution of the species and for the repertoire [of behaviors] acquired by each member." The main point Skinner hoped his readers would take to heart and allow behavioral science to act upon was that "the evolution of a culture is a gigantic exercise in self-control. . . . A scientific view of man offers exciting possibilities."[40]

For Chomsky such a conclusion was politically dangerous and certainly not scientific. He scoffed at Skinner's supposedly "scientific" assumptions, noting, for example: "It is hardly possible to argue that science has advanced only by repudiating hypotheses concerning 'internal states' "—or traditional philosophical and psychological explanations of individual consciousness. By rejecting the experimental study of self-consciousness, Skinner revealed his hostility not only to " 'the nature of scientific inquiry' but even to common engineering practice." Honesty demanded that a scientist investigate "internal states" if they proved to be "the only useful guide to further research." And "by objecting a priori to this reasonable research strategy, Skinner merely condemns his strange variety of 'behavioral science' to continued ineptitude."[41]

If Skinner had troubled to answer formally Chomsky's objections, he might have rejoined that Chomsky never understood why psychological science had originally given up on introspection as science in the era of

*Skinner did not, however, devalue the usefulness of "the literature of freedom" in the past; indeed, that literature "induces people to act." And it had been of great value in removing certain political and economic oppressions and produced a feeling of freedom. But the emphasis of the book was on the larger point that the future could be made better by recognizing and scientifically reshaping institutional controls—which remained, regardless of the feeling of freedom—through the scientific utilization of positive reinforcement rather than political action. One changed the environment scientifically, not through politics, which masked powerful controls that remained and did not take the future into account.[39]

John B. Watson. He also would have certainly argued that Chomsky's own generative grammar required some unseen, internal, innate mechanism—an internal cognitive process before the simplest syntax could appear—as much as behaviorism required the control of observable variables in an environment to shape behavior. Was it not more scientific to work with observable and controllable variables than to defer to some mysterious innate process?

Chomsky relentlessly disparaged Skinnerian science. "Whatever function 'behaviorism' may have served in the past, it has become nothing more than a set of arbitrary restrictions on 'legitimate' theory construction . . . the kind of intellectual shackles that physical scientists would surely not tolerate and that condemns any intellectual pursuit to insignificance."[42] Here he shifted from a critique of Skinner's science to one of his "intellectual shackles" and his own concern about diminishing individual freedom. "There is nothing in Skinner's approach," Chomsky wrote, "which is incompatible with a police state in which rigid laws are enforced by people who are themselves subject to them and the threat of dire punishment hangs over all." If Skinner were an "honest scientist," he would "admit at once that we understand virtually nothing at the level of scientific inquiry with regard to human freedom and dignity." Furthermore, "a person who claims that he has a behavioral technology that will solve the world's problems . . . is required to demonstrate nothing."[43] Skinner later learned from a friend that Chomsky had said privately of *Beyond Freedom and Dignity* that "beyond bed-wetting it's bullshit."[44]

Chomsky had not demolished Skinner's behavioral science, or proved that science needed internal states of self-consciousness, or proved that behavioral science was not incompatible with a "police state." But he scored points when he observed that a science that *claims* it can solve the world's problems has not much but its claims. Skinner remained civil toward Chomsky when occasionally the latter participated in Harvard doctoral examinations for psychology graduate students interested in language. But personal notes revealed his disdain: "Chomsky, like most linguists, *do take themselves seriously.*"[45] That, however, could be said of Skinner and many behaviorists as well; and by the time *Beyond Freedom and Dignity* appeared, both camps had reason to believe they were at the forefront of intellectual life in America. Later Skinner regretted that he did not more aggressively defend his perspective, but in 1971 and 1972 the level of outrage over his book left him somewhat puzzled about just how to respond.

Critiques came from other areas as well. The poet Stephen Spender called the book "fascism without tears."[46] And the novelist Ayn Rand, perhaps the most abusive of all, likened it to "Boris Karloff's embodiment of Frankenstein's monster: a corpse patched with nuts, bolts and screws from the junkyard of philosophy, Darwinism, Positivism, Linguistic Analysis, with some nails by Hume, threads by Russell and glue by the *New York Post.*" She postulated that "the book's voice, like Karloff's, is an emission of inarticulate moaning growls—directed at a special enemy: 'autonomous man.' " Skinner was obsessed with a "hatred of man's mind and virtue . . . reason, achievement, independence, enjoyment, moral pride [and] self-esteem—so intense and consuming a hatred that it consumes itself, and what we read is only its gray ashes, with a few last stinking coals."[47] Autonomous man as the transcending individual was the ideal of Rand's novels, and she maintained an ongoing quarrel with those who attacked self-interest.[48]

Unfavorable reactions came unsolicited from lesser-known persons as well. A professor, declining to give his university affiliation, sent Skinner a negative review from a Pennsylvania newspaper. He declared that "after almost a lifetime of university teaching I feel qualified to state that virtually every non-psychologist scholar on the university campus considers psychologists nuts. Your newest book substantiates this thought."[49]

The enclosed review found nothing forward-looking or future-oriented in the "beyond" of Skinner's book title. It noted "that Dr. Skinner is [not] ahead of his time. He is behind the times 35 years. He'd have flourished and been honored as a prophet in the Germany or Italy of the 1930s." Indeed, "when individual freedom and dignity cease to matter tyranny takes over. . . . Ask the ghosts of Dachau. Ask the spirits which haunt the bleak vastness of Siberia." The editorialist made an impassioned defense of the representatives of the hallowed American tradition Skinner had maligned: "In his role as cosmic puppeteer, Dr. Skinner professes to see 'signs of emotional instability in those who are affected by the literature of freedom.' Thomas Jefferson instantly comes to mind. Abraham Lincoln. Oliver Wendell Holmes. Unstable? If so let us seek out and develop to its full richness that sort of instability."[50] Reflecting upon this abundance of vituperation, the biophysicist John

Platt declared that "Skinner may have had the worst press since Darwin."[51]

The massive criticism of *Beyond Freedom and Dignity* by American intellectuals, writers, academics, and editorialists and just plain enraged citizens showed clearly that Skinner had done more than touch a sensitive cultural nerve; he had trespassed upon, violated, the basic national creed.[52] The extraordinary turmoil of the late 1960s and early 1970s helped exaggerate the transgression. Political assassination, racial violence, an increasingly unpopular war, and what many considered to be indecent campus life-styles and protests dominated both written and electronic media, supercharging public emotions. Given this volatile milieu, it is necessary to look historically at the tradition Skinner's critics claimed he discredited.

BY THE EARLY NINETEENTH CENTURY Americans had divided into two camps, two national creeds with significantly different emphases—"a split in culture between two polarized parties: 'the party of the Past and the party of the Future,' . . . or the 'parties of Memory and Hope.' " Before the Civil War these two "parties" engaged in a fruitful national dialogue in politics and literature. But since that great conflict one party had emerged triumphant. The party of Hope or, as one might call them, the liberators, was in command. Their intellectual and social victory over the party of Memory, or what one might call the traditionalists, seemed complete.[53]

The crux of the issue between the two groups was the relationship of the individual to social institutions, the degree of social control necessary in a democratic nation. The liberators insisted that institutions and, by implication, society, corrupted the individual. Ideally, the free individual was in a position to make correct moral choices and in time, one by one, could effectively reform, and even perfect, society. But the traditionalists, following the heritage of the New England Puritans and the Federalist party, argued that history showed the free individual to be a menace to social well-being. Only if individuals were restrained within institutional boundaries—limits set by church and/or government—could a morally responsible and beneficent society be maintained.

By the mid-nineteenth century this vital debate had dissolved. Most Americans living in the North extolled the virtues of an unbounded, freewheeling individualism, while white Southerners were incensed over northern trangressions against liberties won in the era of the American Revolution: state sovereignty and the right to control their property, especially black slaves. The Civil War was a conflict not between liberation and tradition but between two different interpretations of freedom: "Both sides in the American Civil War professed to be fighting for freedom."[54] *Freedom* became the most provocative and beloved word in national discourse; almost everybody was for it, and many had died for it. The triumph of the North was far more than victory over the South and the reestablishment of the Union and the end of slavery; it reaffirmed a national creed that emphasized individual liberty, a creed that Northerners and Southerners alike could embrace as the essence of being American.

By the early twentieth century American individualism was defended more and more as a birthright, indivisible from the meaning of freedom, an indelible and sacred national heritage—a *creed* largely untouched by the periodic efforts of intellectuals and politicians to initiate reforms that promoted social welfare and responsibility. Two horrific world wars, a devastating economic depression, and looming nuclear, ecological, and population catastrophes did not diminish liberator dominance. Indeed, when *Beyond Freedom and Dignity* appeared, the appeal of American individualism and its linkage with the heritage of freedom were stronger than ever.[55]

Skinner has been most often attacked for being against the liberator heritage that put individualized expectations of equality of opportunity, social justice, and the good life at the forefront of the American creed; rarely has he been maligned as a defender of individualism. And even then, the kind of individualism Skinner supported was in reality a ploy to control, a veiled attempt to make people behave in old-fashioned quasi-Fascist ways. For example, the sociologist Richard Sennett accused Skinner of having a "hidden agenda" in *Beyond Freedom and Dignity,* the purpose of which was to defend "the articles of faith in Nixonian America, of the small-town businessman who feels life has degenerated, has gotten beyond his control and who thinks things will get better when other people learn how to behave."[56] Additionally, he was criticized as an enemy of the individual's creative role in society. As Joseph Wood Krutch had remarked in the early 1950s, Skinner "for all practical purposes [makes] man . . . merely the product of society."[57] And true to form, Carl Rogers, in 1956, echoed Krutch by approvingly quoting John Dewey, one of the most visible twentieth-century liberators: " 'Science has made its way by releasing, not by suppressing, the elements of variation, of novel creation in individuals.' "[58]

If Skinner is placed in the heritage of discourse between liberators and traditionalists in the making of the American creed, the negative reactions to his social thinking, especially to *Beyond Freedom and Dignity,* become more understandable.[59] By the same measure Skinner's own social thinking becomes distinctive, even unique. His book, especially its title, reminded Americans that the traditionalist perspective, now operating under the guise of behavioral science, had not yet been thoroughly routed. Skinner seemed to support reconsideration of the essential ingredient in the traditionalist perspective—the older view, the Puritan-Federalist perspective that emphasized efforts to sew into the fabric of New World society, not free individuals, but civic responsibility and the sacrifice of individual interests and liberties to society.[60] Skinner seemed to be asking Americans to dismiss the liberated individual and thereby return to an American past in which the individual was forced to behave in ways that benefited institutions. Thus behavioral science seemed simply another oppressor—an instrument of control, like churches or governments or political parties, which argued for social good but in effect threatened or destroyed individual liberties.

Skinner never wanted to return to a society in which social responsibility was maintained through aversive control. Yet control had never been and could never be absent: "We all control, and we are all controlled. As human behavior is further analyzed, control will become more effective. Sooner or later the problems must be faced."[61] Inexorably bound to the problem of control, however, was the crucial matter of building social responsibility into the future. "The important thing," Skinner noted, "is that institutions last longer than individuals and arrange contingencies which take a reasonably remote future into account." Indeed, death was "the final assault on freedom and dignity, the last affront to the individual." The staunch individualist could

> find no solace in reflecting upon any contribution which will survive him. He has refused to act for the good of others and is therefore not reinforced by the fact that others whom he has helped will outlive him. He has refused to be concerned for the survival of his culture and is not reinforced by the fact that the culture will long survive him. In defense of his own dignity he has denied

> the contributions of the past and must therefore relinquish all claims upon the future.[62]

Yet to the liberators, Skinner's case for a social future controlled by behavioral scientists was easily associated with individuals and groups who manipulated power—a theme that has preoccupied and at times obsessed many Americans since the birth of their nationality.

Well before the publication of *Beyond Freedom and Dignity,* Carl Rogers expressed a concern about Skinner and behavioral scientists that echoed an American heritage of suspicion about who holds power and how power is used. During their 1956 debate, Rogers struck at the heart of the liberator concern when he remarked, "I believe that in Skinner's presentation here and in his previous writings, there is a serious underestimation of the problem of power." He contrasted his own seasoned understanding of recent history against Skinner's historical naïveté:

> To hope that the power which is being made available by the behavioral sciences will be exercised by the scientists . . . seems to be a hope little supported by either recent or distant history. It seems far more likely that behavioral scientists, holding their present attitudes, will be in the position of German rocket scientists specializing in guided missiles. First they worked devotedly for Hitler to destroy the U.S.S.R. and the United States. Now . . . they work devotedly for the U.S.S.R. in the interest of destroying the United States, or devotedly for the United States in the interest of destroying the U.S.S.R. If behavioral scientists are concerned solely with advancing their science, it seems most probable that they will serve the purposes of whatever individual or group has the power.[63]

Rogers's concern about government scientific manipulation made as much sense to American liberators in the early 1970s as it did during the cold war.

Skinner, however, was not put off by the question of power and who had it. The major issue was not *who* had the power but *how* society was to be shaped. " 'Who *should* control?' " Skinner asserted, "is a spurious question. . . . If we look to the long-term effect upon the group the question becomes, 'Who should control if the culture is to survive?' " This was the equivalent of asking, "Who *will* control in the group which does survive?"[64] Hence it was the survival of the group—that is, "culture"—that counted most. But by insisting in *Beyond Freedom and Dignity* that behavioral scientists could best engineer cultural practices that ensured human survival, Skinner triggered the long-standing American concern about the use of power to shape social destiny. His talk of "control" was easily associated with the abuse of power. Indeed, he did it himself: "To prevent the misuse of controlling power . . . we must look not at the controller himself but at the contingencies under which he engages in control."[65]

Skinner tried to defuse the concern about control, but he never fully appreciated why the word evoked such outrage. A creed emphasizing individual freedom had brought misunderstanding about the nature of control. "We [the American public] do not oppose all forms of control because it is 'human nature' to do so," he explained. Such a presumption was "an attitude which has been carefully engineered in large part by what we call the literature of democracy." Again and again he insisted that control was inevitable, whatever its disguises: "Through a masterful piece of misrepresentation, the illusion is fostered that . . . [free institutions] do not involve the control of behavior." But "analysis reveals not only the presence of well-defined behavioral processes, it demonstrates a kind of control no less inexorable, though in some ways more acceptable, than the bully's threat of force."[66] Everyone is controlled, whether by peer pressure, the need for approval, advertisements, or fear of punishment. Skinner wanted to replace ineffective and harmful means of control with scientifically designed mechanisms of self-control: environmental contingencies that would positively reinforce individual behaviors, which, when conditioned, *automatically* produce a socially responsible culture.

TO MOST READERS of *Beyond Freedom and Dignity,* such scientific engineering was as reprehensible as any other form of control. They understood Skinner and his behavioral science as an anachronism: the old story wrapped in the new terminology of someone claiming to know what was good for the public but really interested in manipulating power at the expense of the free individual.

Skinner was involved in a delicate balancing act. On the one hand, he recognized the importance of the evolution of Western democracy, the historic fight against political tyranny, the growth of the freedom of the individual. He valued the *feeling* of freedom. On the other hand, he recognized the *illusion* of freedom and the possibility of scientifically controlling environmental contingencies to produce individuals who practiced behaviors that were socially beneficial. Future social progress, he emphasized, rested not with those who quested for freedom of the individual but with the behavioral scientist who continued humanity's protracted struggle to control nature and itself.

The United States has had a protracted history of deep ambivalence about the relationship of "freedom" to "control." The Puritans wanted to free themselves from the corruptions of the Church of England by establishing intellectual and social control in New England. The revolutionary generation wanted to free itself from the tyranny of the British parliament and build a "republican" society that controlled individual ambitions for the good of the nation. The South tried to free itself from the Union to preserve and control its slave society. After the Civil War, Americans sought to build an industrialized nation—to "order" society—without giving up individual freedoms. Twentieth-century Americans have embraced a powerful controlling technology as a boon to personal freedom. The canvas of American history has been painted with large and small motifs representing recurrent difficulties in defining a stable relationship between freedom and control, between liberation and tradition, between the individual and society.[67]

Skinner wanted behavioral science to free America from this historical predicament. One of the few favorable reviews of *Beyond Freedom and Dignity* suggested that the counterculture upheaval of the late 1960s and early 1970s was as much a search for control as for freedom. Skinner reminded us "that human beings need not only freedom but mutual control as well. . . . Maybe this is what our alienated youth are asking for: the right to control others . . . in return for being controlled to some extent." The review concluded with a statement Skinner would have enthusiastically endorsed: "Revolutions should be about the business of replacing aversive controls . . . with positive fulfilling ones."[68] That was not only a major emphasis of his book but the social objective of behavioral science.

Beyond Freedom and Dignity presented a novel synthesis of the liberator and traditionalist heritages in American history. The methods of behavioral science were applied to the individual—whether rat, pigeon, or human being. Behavioral engineering—"the design of culture"—was a matter of positively reinforcing the individual so that the environmental contingencies could support cultural practices that guaranteed a livable future for everyone. Skinner worked within the liberator, party of Hope perspective by trying to remove aversive human conditions and replace them with positive-reinforcing ones. In this sense Skinner was a liberator. Indeed, his creations as a social inventor were meant to be liberating: The aircrib liberated the child from physical restraint and the mother from domestic drudgery; *Walden Two* liberated the modern individual from a wasteful, competitive, unrewarding world; and a technology of teaching liberated teachers from custodial and disciplinary roles in overfilled classrooms.

Skinner's objective, however, was not to "free" humans from social controls; that was not possible, nor did the illusion of being a "free individual" promote human survival. Rather, by shaping the environmental variables of which individual behavior is a function—or, more generally, by careful environmental design—the individual would become socially responsible, an absolutely essential behavior for a livable future. But with the shaping effects of a technology of positive reinforcement, the individual did not have to depend on a personal conscience, God, or a social cause to be a good citizen; one did not do one's civic duty for fear of punishing consequences, whether personal or social. Hence Skinner rejected the usual American practices that have historically promoted social responsibility. Not a personal conscience or a social crusade or governmental restraints or religious rewards would effectively promote the survival of a responsible social culture into the twenty-first century. Skinner stepped outside both the party of Hope and the party of Memory, outside liberation and tradition, to present Americans with another cultural alternative.

American history in a fundamental sense has been the saga of transcending restraints (Skinner would have said "aversiveness"); and that, in turn, is the story of a national environment that shaped an individualistic, democratic, and technologically oriented culture. Skinner was an articulate and innovative example of that native tradition. His legacy as a social inventor, however, is not simply that of an interesting cultural artifact. Few, if any, twentieth-century Americans have so tellingly translated and challenged our social predicament. For if the individual and the environment are the great focuses of the American experience, then Skinner urged us not only to consider them in relation to each other but to consider them *scientifically. Beyond Freedom and Dignity* argued that behavioral science had enormous species-saving application.

Skinner asked Americans to see the radical potential in the American experience, to consider it as a scientific application of self-control for social survival—to face the *possibility* of fashioning a better society without relying on either the liberator or the traditionalist heritage. And it may well be the very novelty of his argument that shaped the reaction of his critics.

10

Master of Self-Management

I am the person I'm most concerned with controlling.

—"Will Success Spoil B. F. Skinner?" *Psychology Today* (1972)

Skinner never again captured the intense media attention of those frenetic months of late 1971, but neither did he drift into public obscurity. Newspaper and magazine features appeared periodically, and he made an occasional television appearance: "Donahue," "Good Morning America," and a 1979 PBS "Nova" special called "A World of Difference: B. F. Skinner and the Good Life," which focused on his Susquehanna boyhood, his science, and Twin Oaks, the Walden Two community. His books continued to sell well, especially *Beyond Freedom and Dignity,* which appeared in translation from Japan to Brazil to Sweden. Although he often vowed he would reduce his professional involvement with behaviorist organizations such as Division 25 of the American Psychological Association (APA) and the Association for Behavioral Analysis (ABA), he continued to attend conferences and often delivered papers. As the years passed he became revered as a legendary figure in American psychology. In 1970 *American Psychologist* listed Skinner second, after Freud, in influence on twentieth-century psychology.[1] And in 1989, after viewing a bust of himself at the Cambridge Center for Behavioral Studies, he mentioned to a friend that his name was now cited in psychological literature more often than Freud's. " 'Was that an objective?' " the friend asked. Skinner replied, " 'I thought I might make it.' "[2]

But being a legend in his own time had its drawbacks. Everyone knew Skinner, but relatively few treated him as a friend. As a free-lance writer who met him at a conference put it, "he is never asked out for a beer with the boys at conventions or included in dinner plans. That's how I met him; we were together at a convention and neither of us had been invited by any group to lunch. He, because nobody thought the great B. F. would go, and me because I knew nobody. So we lunched together, and he told me it happened all the time."[3]

He especially enjoyed upstaging the more conventional professionals, not with his reputation but with humor, charade, and dramatic flair. He once appeared at a convention dressed as Pavlov and once as a fundamentalist minister. He even staged a verbal fight with his old friend from the early Harvard years, Fred Keller, with whom in fifty years of friendship never an angry word had been uttered.[4] Indeed, behaviorists who attended APA and ABA conventions expected some kind of Skinner-Keller fun whenever the two were on hand. Yet Skinner disliked organizational politicking, as well as the increasing cognitive leanings in both these organizations. He consistently declined invitations to serve as president of the APA, once remarking, "even though the Presidency of the APA is no doubt a great honor, it is also a hell of a lot of work, and a kind of work at which I am no good."[5] Nevertheless, at the end of his life he regretted that he had not been more of an "empire builder." Young behaviorists were having difficulty getting academic jobs in a profession that was more and more dominated by cognitive psychology, and he felt partially responsible for their competitive disadvantage.[6]

Skinner also continued to find great joy in his daughters. Julie had married in 1962. She and her husband, Ernest Vargas, had both become dedicated behaviorists, teaching at the University of West Virginia and recently helping found a new behaviorist organization, The International Behaviorology Association (TIBA), as an alternative to nonbehaviorist trends in Division 25 and the ABA.[7] The Vargases also made Skinner the grandfather of two girls—Lisa in 1966 and Justine in 1970—both raised in aircribs and adored and observed by Skinner with the same acute attention he had given his own daughters.[8] In the 1960s Deborah had done behavioral work with porpoises in Hawaii, but she did not, as Julie did, follow her father into a professional career as a behaviorist. She lived in London after her marriage in 1973 to Barry Buzan, an economist at the University of London, and the couple remained childless. Deborah pursued art, producing accomplished etchings, and wrote articles on London restaurants.

Eve and Fred continued their long-lived marriage at the Old Dee Road residence in Cambridge. He continued to sleep alone, an arrangement partly necessitated because he was an early riser and she a late one. They found, however, an equilibrium if not a happiness that had earlier eluded them, in part, because of Eve's career unfulfillment. In 1976 she began reading once a week for the Braille Press, making tapes of books that blind students requested for their courses. More important, she found a career in the 1970s as an art history guide/instructor at the Boston Museum of Fine Arts. The role required intensive training—reading and research in art history—in order to become expert at presenting information to museum visitors. Eve loved her work and continued this late-in-life career into the 1990s. Julie recalled an occasion when her mother glowed with pride when Fred attended an event at the museum and was introduced as the husband of Eve Skinner.[9]

Skinner claimed shortly before his death that he had tried very hard to maintain a pleasant environment for Eve, even though she had not taken much interest in his work or his kudos. He clearly regretted her lack of interest.[10] But Eve helped him behind the scenes, preparing lunches (she became an accomplished cook) and seeing to it that the house, if not his study, stayed spotlessly clean. Although they made occasional trips to Europe, especially to London to see Deborah and Barry, both enjoyed living in Cambridge more than anywhere else; and it was the Cambridge-Boston environment, a few close friends, and the opportunity to pursue fulfilling interests that stabilized their relationship.

On the surface, Skinner's later years seemed conventional: He retired from Harvard in 1974 and was presented with a first edition of Thoreau's *Walden.*[11] He lived quietly with his wife and enjoyed visits from his grandchildren. One family birthday party featured a cake with the inscription: "For reinforcement Blow Out the Candles."[12] Like many retired faculty members who had attained some reputation, Skinner kept an office—although he paid for his secretary's work—and spent several weekday hours working on correspondence and seeing visitors. He continued to walk to the William James Hall office until a friend reported to Eve that she had seen him nearly struck by an automobile. On summer afternoons, although Skinner no longer swam, he enjoyed the company of friends by the backyard pool, and a vodka tonic or two. It was the life-style of an upper-middle-class American who now had the time and affluence to relax and enjoy a well-deserved retirement.

On closer inspection, however, he had continued the intensity of his intellectual life. Skinner never retired from the task of "discovering what I have to say." Shortly after finishing *Beyond Freedom and Dignity* he made plans for four more books: "an autobiography, a notebook, a primer of behaviorism and a little book on how to think."[13] The circadian rhythm of his writing schedule did not miss a beat. A timer rang at midnight signaling him to arise, move to his desk, and write until signaled to stop at one o'clock. He returned to bed and arose again to the sound of the timer at five o'clock and composed until it buzzed two hours later. He wrote three hours a day, seven days a week, holidays included. As the years passed, these three early morning hours were, as he often said, "the most reinforcing part of my day."[14] The other twenty-one hours were arranged to make the writing time as profitable as possible.

During the mid-1980s Japanese friends had sent him a *beddoe,* what Skinner referred to as "my cubicle," which he had installed in his basement study. It was an enclosed sleeping space, made of bright yellow plastic and modified with the installation of a timer and a stereo system. Under the sleeping pallet was a drawer containing music cassettes, many of which were of his favorite composer, Richard Wagner. During the last decade of his life he frequently listened to Wagnerian music, usually in mid-afternoon, relaxing after the rigors of early morning writing and thinking. Because his hearing had deteriorated and because the volume at which he liked to listen bothered others in the house, he used earphones. But sometimes the strains of Parsifal or one of the Ring operas would drift through the basement.

Music had been important to him all his life. During his boyhood, he had been much affected by his mother's playing and singing at the piano in their Susquehanna home. He had played the saxophone in high school, and at Hamilton College had traveled with the college chorus and enjoyed musical ensembles at the Saunderses' home. He had purchased a small piano during his second year at Harvard, attended concerts, and listened to classical recordings on his own phonograph. He had often played piano with other musically inclined colleagues at Minnesota. And for decades, while a professor at Harvard—until arthritis in his fingers prevented it—he played a clavichord for half an hour every day to relax. (The clavichord still remains in the foyer of the Old Dee Road house.) In the 1970s he even had a small organ in his basement study.

But music was more than an enjoyable pastime; it was vitally related to Skinner's creative life, to discovering what he wanted to say. Wagner, especially, produced a mood that visual media could not begin to approach: "I feel very uplifted by it all. It has that ennobling effect. It is movement. It is time."[15] Wagnerian opera evoked a rich impression of experience, displaying "human behavior in all its aspects."[16] He kept a notebook at hand while listening so that he could immediately record a significant thought.

Unlike many other American intellectuals, such as Emerson, Thoreau, and William James, Skinner did not seek inspiration by communing with nature; he did not regularly retreat to the solitude of the woods or the shore to escape society and rekindle his imagination. When he owned the cottage on Monhegan Island, he "would often get up and go and sit on the rocks and watch the waves. But I don't think I was communing with anything." And while at Putney, Vermont, in 1955 he had "walked a good deal and communed with myself there but not with nature."[17] Music, the solitude of his study, and a schedule of writing were what he needed to maintain his intellectual productivity. And there was an even more cloistered environment he needed: "I used to build boxes to sit in and so on," he recalled. "I sleep in a box. I put Julie's children in boxes. It is a kind of isolation, but it is an isolation in which it is possible to have more or less ideal conditions—ideal for the child and baby. Ideal for me." In his *beddoe,* enclosed in what at first glance appeared to be a yellow submarine, Skinner regenerated himself, with rest and music, for the final intellectual work of his life. This was, however, "very different from escaping."[18]

HE PRODUCED a considerable amount of work during the last two decades of his life. Even before finishing *Beyond Freedom and Dignity,* he had begun working on an autobiography. It appeared in three volumes: *Particulars of My Life* in 1974, *The Shaping of a Behaviorist* in 1979, and *A Matter of Consequences* in 1984. The autobiography was by far the longest published work of Skinner's career, running over 1,300 pages. He tried to write from "the record" rather than "from memory." He explained that he had written it so that "people would like me." He wanted them to like him in the sense of having affection for him, but also in the sense of "becoming more like me."[19] He also wrote in an objective style, as if he were "the other," and scrupulously avoided discussing his feelings. He wanted it to be "the autobiography of a nonperson," the story of the effects that personal and cultural history had had in shaping his life and work. He would take no credit, make no attempt to maintain that he had special genius or inner strength: "By tracing what I have done to my environmental history rather than assigning it to a mysterious creative process, I have relinquished all chance of being called a Great thinker."[20] He translated himself and his autobiography into behavioral terms: "I am the locus in which certain genetic and environmental forces have come together to produce my behavior. I am very much interested in looking into my past . . . to find out exactly, as far as I can, what sort of conditions are indeed responsible for the things I've done."[21]

But if Skinner had written an autobiography about the way he felt—even from memory, with all the distortions and illusions such an approach entails—he might have convinced more people both to like him and to be like him. The objective tone of his autobiography was not only behavioristic in the sense that it simply presented a record of what had happened to him; it was also distantly related to his desire to become an "objective writer" during his Dark Year in Scranton. And after writing *Walden Two* he had occasionally expressed a desire to write another novel. He told his friend Silvia Saunders (Percy's daughter) that "I have often wondered about the possibility of writing a novel which presents the facts as life presents them, leaving it to the readers to respond emotionally as life leads them to respond." He would not "be trying to tell the readers what to feel or what his characters felt. . . . I decided to try it out in the autobiography."[22] Her reaction, however, to Skinner's effort to be objective about his life was to wonder whether he might be "a person of not very deep feelings."[23] Skinner never denied that feelings existed—in fact, he was at times quite sentimental. But he saw that "deep feelings" were evoked by something that happened in the environment, verbal or otherwise; they were not themselves the cause of emotions.

He also wrote a "primer," *About Behaviorism,* which appeared in 1974. This book was purposely designed to correct misconceptions that intelligent nonbehaviorists held about behavioral science. It was a nontechnical course in what behaviorism was and was not, offered in the hope that if the educated public knew what behaviorism really was, in language they understood, they might accept it or at least be less antagonistic toward it.

In the introduction Skinner listed twenty misconceptions commonly made about behaviorism, such as that "it ignores consciousness, feelings and states of mind," and that "it dehumanizes man . . . is reductionistic and destroys man *qua* man."[24] Clearly explain the mistakes commonly made about behaviorism, Skinner maintained, and people would cease to make the mistakes. Reviews were relatively favorable, one reviewer calling the book "a much more effective work than *Beyond Freedom and Dignity* precisely because it takes more seriously the tradition . . . of its opposition."[25] Yet *About Behaviorism* did not receive nearly as much attention as its infamous predecessor. One reviewer erroneously asserted that "the battle over Skinner's ideas is just beginning." But she was certainly correct in calling it "one of the most interesting contests of our generation."[26]

In fact, the battle had climaxed with the immediate reaction to *Beyond Freedom and Dignity.* The American reading public had made that book an instant best-seller, but had just as surely rejected Skinner's argument that there were cultural matters more important than preserving and extending individual freedom. *About Behaviorism* did not turn things around. Popularity and infamy still coexisted. Skinner noted to a friend that after being cheered by "an audience of six thousand students at the University of Michigan last week," he was "hanged in effigy at Indiana University the same day."[27] He lectured at Reed College, where two students in the audience were dressed as rats.[28] Skinner enjoyed the joke. The Humanist Society named him "Humanist of the Year" for 1972. And the Thoreau Society invited him to speak on "Walden (One) and Walden Two," an occasion vindicating his conviction that he had "always been a Thoreauvian."[29]

During the late 1970s and 1980s Skinner gathered together his continuing outpouring of lectures and articles into three books: *Reflections on Behaviorism and Society* (1978), *Upon Further Reflection* (1987), and *Recent Issues in the Analysis of Behavior* (1989).[30] These collections were intended for behaviorists and psychologists as well as the general reader. They addressed broad social questions such as, "What is wrong with daily life in the Western World?" and, "Why are we not being more active about the future?"[31] He also answered over 140 commentaries on six crucial articles he had previously published on the theoretical aspects of behaviorism and behavioral analysis, the "Canonical Papers of B. F. Skinner." Articles, critical responses, and Skinner's short answers to the commentaries appeared as *The Selection of Behavior, the Operant Behaviorism of B. F. Skinner:*

Comments and Consequences (1989).[32] At the very end of his life he was working on a consideration of ethics from both Darwinian and operant perspectives. He intended to write the book without reference to moral authorities and would not worry about whether or not the lay public understood it—a reversal from his efforts to reach the public in *About Behaviorism.* As he aged he increasingly realized that the popularity of his science was beyond his control.[33]

HE ALSO FOUND it more difficult to grasp new thoughts and think his work through as a whole. A major emphasis of his adult life had been on managing his work schedule so as to ensure intellectual production. He not only started and stopped writing by the buzz of a timer; he counted the number of words he had written in a given period, kept records of how many hours he spent on each book, and made a graph charting the sales of *Walden Two*—a very reinforcing activity, as sales approached two million in the late 1980s. Efforts to arrange a working environment to effectively mitigate or deflect the effects of age became a central focus of his life. Skinner, with a younger colleague, even published a small book on the subject.[34]

After finishing *Beyond Freedom and Dignity* at age sixty-seven, however, he was especially exhausted. He had felt symptoms of angina, and was told by his physician that one-third of those with his condition—clogged arteries—did not survive beyond five years. Eve and Julie put him on a strict diet to lower his cholesterol. The uproar over his book had disrupted his working schedule and left him feeling that he had behaved inappropriately: "At this moment I feel as if I had betrayed some sacred trust during the past year. I have been unduly affected by my book and all the publicity it has received. . . . I want to get back to work, but the 'purity' of my motives is suspect. I am defending myself, for example. I must hold to my schedule and my plans if I am to save myself." His resolve to return to a schedule of putting intellectual efforts first was not only "to 'offset a sense of failure [and] discouragement' " but "to offset extinction," meaning ceasing to behave as an intellectual rather than a public figure defending his creations.[35]

Skinner was occasionally distressed, for example, to find that a seemingly new idea in a note or draft for a paper had appeared in something he had published years earlier. He worried about plagiarizing himself.[36]

His efforts between 1972 and 1990 to achieve intellectual self-management were an especially interesting application of behavioral technology to himself. Richard Herrnstein, a younger colleague at Harvard who had intellectually parted ways with Skinner, continued to admire his successful self-management in the face of mounting personal difficulties.[37]

By the mid-1970s, although his general health remained good, he had lost much of his hearing. In addition to wearing a hearing aid, he devised in his basement workshop an audio system utilizing "Mickey Mouse ears," which better amplified sound waves, so he could continue to listen to music. In 1973 he began to experience eye troubles. By 1977 his left eye was useless for reading because of a distorted macula due to a small ruptured blood vessel. His right eye developed glaucoma, and he lost almost half his field of vision. Fearing he was going blind, Skinner wrote a physician for advice about the use of marijuana to relieve the suffering of glaucoma. He decided against this therapy, however, because of risks to those with a history of heart problems.[38]

But there were other ways to counteract his debility. He purchased all the symphonies of Mozart and the organ works of Bach, intending "to specialize in music if I could no longer read." Writing to a man whose wife was losing her sight, Skinner outlined a behaviorist program to "shift to non-visual stimuli while continuing to follow the same interests":

> The main thing is programming a person into these activities. It is too much to try to do them all at once. That produces extinction and a sense of frustration. Little things first so that reinforcement can be frequent. I am just about to start something along these lines. I have a small machine shop in my house but can no longer do fine work. I have decided on gardening under artificial light as something which will be pleasing but require very little skilled movement. I am going to begin very slowly hoping that I will be reinforced by the results. Only then will I expand to a larger scale.

Because any sign of progress was likely to be reinforcing, Skinner advised his correspondent's wife to keep a cassette diary, recording events of her day, to which she could return "later to see the extent of her progress."[39]

He successfully engineered an environment in which he could still be reinforcingly productive. His sight and hearing remained poor, but until the late 1980s no other physical ailment threatened to derail his writing schedule. He used cassettes and fashioned a magnifying lens on movable metal arms attached to the chair in his study so he could continue to read, if slowly. He invented a Thinking Aid, a system of filing for his basement study desk that allowed him to create large lettered outlines, later to be filled in as he gradually discovered what he wanted to say.

Then in 1981 a cancerous lesion was found in his head. "I did not miss a heartbeat when Eve told me the tumor removed from my head was malignant," he remarked.[40] He began radiation therapy immediately. Although able to accept without undo anxiety threats to his health, Skinner was never fatalistic in the sense that he continued to behave as if nothing had happened. Freud, for example, continued to smoke cigars after being diagnosed with cancer of the mouth. Skinner's science shaped his reaction to personal calamity, and his reaction was to do something beneficial to his environment that would reshape his behavior in profitable ways.

In December 1987 he took a fall down the basement stairs leading to his study, struck his head, and suffered a life-threatening subdural hematoma. Twice he underwent surgery to relieve pressure on the brain, each time being under anesthesia for several hours. Although by mid-1988 he was "back on schedule"—writing in the early morning hours and later going to his William James Hall office, the trauma affected his intellectual performance: "I made many mistakes when writing. I would write the wrong word. It was often hard to execute the writing of a word. . . . After the second operation I had episodes when I could not write at all." Yet he continued to work on his ethics book, using his Thinking Aid outline even more rigorously.[41]

Then, in early November 1989 while at the hospital where Eve was being treated for an embolism, Skinner fainted. Doctors admitted him, took blood samples, and found him markedly anemic, a serious but treatable condition. To be thorough, however, they did a full hematology. The next morning two grim-faced physicians came into his room. Skinner recalled saying before they could speak, "You're going to tell me I have cancer, aren't you? Well, how much time do I have?" The doctors estimated six months to a year. But he insisted he did not have the "slightest anxiety" about his approaching death. His only concern was the effect his demise would have on his daughters, especially Deborah, who "had cried half a day when the family cat died."[42] When the news of his fatal illness spread, he received many letters of support, some with suggestions for treatment. One letter arrived from England with a bottle of flaxseed oil, with a recommended dosage of two tablespoons a day.[43]

The doctors assured him that he would be able to function fairly normally right up to days, even hours, before he would lapse into a coma and die. And he did. But during the ten months Skinner lived with leukemia, he had to endure some unpleasant medical procedures. Every eight days he needed a complete blood transfusion, which required eleven hours to complete. Every four days he needed platelets, another four- or five-hour ordeal. Julie had the stressful task of injecting medicine directly into his vein. While at the hospital he would listen to reinforcing music on a compact disk with earphones, and he kept a notebook handy. He maintained his early morning writing schedule, though he stopped going to the office because of dangers of infection. He continued to have visitors regularly, but they were requested to wash their hands before seeing him. A secretary set up an office in the basement next to his study to help him answer his correspondence and type his handwritten notes and papers. Eight days before his death on August 18, 1990, at the age of eighty-six, Skinner delivered a twenty-minute speech at the APA convention in Boston without notes. Weighing 120 pounds and looking frail, he addressed, with a strong and resonant voice, a standing-room audience for the last time.

SKINNER WOULD HAVE DENIED THAT his remarkable perseverance was either heroic or courageous, resulting from great inner strength or will power. His intellectual life had been maintained by careful self-management, by arranging his environment to encourage the conduct he wanted to perpetuate. He viewed himself as the organism, shaped by the effects of environmental contingencies, he knew best. He viewed his own death with the same detachment as he had viewed his Grandfather Burrhus's demise sixty-four years earlier: "I believe that what will happen will be that my body no longer works and that's the end of the body and the end of me as a person, because I need the body to be a person."[44]

As he practiced the self-management necessary to continue his various writing projects, several focuses emerged. He was increasingly concerned about cognitive psychology, whose professionally dominating presence he viewed with alarm. On one level, he worried about its effect on the professional future of behavioral analysis. "I have often felt the lack of young operant conditioners interested in and capable of writing about current problems," he wrote in 1974. The more verbal operant students were drawn to "information theory, cognitive psychology, humanistic psychology, etc," failing to find academic positions unless they did so.[45] He regretted the susceptibility of American psychology to fads, and worried about the job opportunities for those trained as behaviorists. Although there were universities such as Arizona State, Kansas, and Western Michigan with strong behavioral departments, the top institutions such as Harvard and Yale and Columbia had embraced the cognitive trend.

But what worried Skinner most was that cognitive science—and he often lumped it along with "information theory" and psychoanalysis into "mentalism"—was no science at all. During the late nineteenth century, when psychology had first attempted to become scientific, the basic error had been to attempt to derive the dimensions of mental states from physiology—from sensations, nerves, and the like. In America Edward Bradford Titchener at Cornell and William James at Harvard had attempted to develop, respectively, "structural" and "functional" sciences of consciousness.[46] James, especially, had appealed to physiology when discussing mental states. An emotion such as sadness, for instance, was caused by the physical sensation of crying. Now, despite the work of early behaviorists such as John B. Watson, who had shown the impossibility of establishing a science of consciousness, the cognitive psychologists were again talking about mental states derived from the brain. It was as if they had retrogressed the better part of a century.[47]

Skinner believed that no contemporary cognitive psychologist had gotten much further than William James. And "to say that we shall eventually find the physical dimensions [of the mind] *via* physiology is not to justify that practice." He predicted that "what will be found by physiology are not the physical correlates of the mind but the physical conditions felt."[48] By the late 1980s the cognitivists were appealing more and more to the authority of brain science. But the brain was part of the body and, once that was admitted, one was back to talking about behavior. He certainly did not deny that physiology was a science; that, however, was different than claiming that consciousness could be derived from physiology.

He was once asked, If mentalism is really so powerless, why has it held the field so long? and replied:

> I believe I gave a correct answer by pointing to evolution. My questioner might have asked Darwin, "If natural selection is so powerful, why have people believed so long in the creation of the species according to Genesis?" The myths that explain the origin of the universe and the existence of living things, especially man, have been extremely powerful and are not yet displaced by a scientific view. Mind is a myth, with all the power of myths.[49]

Skinner repeated this pronouncement on the mythical as opposed to the scientific power of "mentalism" in his last APA address, suggesting that the creationist resistance to natural selection was paralleled by the contemporary resistance of cognitive psychologists to behavioral analysis. There were gasps of disbelief and dismay, as well as approval, from his audience.

The magnitude of his scientific differences with the cognitivists, indeed with all nonbehaviorist psychology, was recorded in a 1977 note written from the perspective of the history of science:

> People like William James and Freud extract principles from human behavior and use them to explain similar instances of human behavior. That is what astronomy did before there was a science of physics. There were principles peculiar to the heavens—some order, some usefulness within the solar system only.
>
> I get my principles under controlled conditions as astronomers get their principles from the physics laboratory. And I can *interpret* human behavior as they can interpret the radiation that comes to them from space. The principles are not derived from the facts they are used to interpret.[50]

Behavioral analysis, like modern physics, is science. Most psychology is not science. In this sense, Skinner had never been a psychologist. Implicit in his distinction between psychology and science was a fear for the future of science in general, especially as a practical technology to solve human problems and provide for a future. "The whole culture is suffering from the rejection of science and may go down the drain as a result," he remarked in the 1970s.[51] "Talking about feelings is safe because nothing important will ever be done about them. . . . Talking about changes in the social environment is dangerous stuff."[52] As Skinner aged he became more pessimistic about the ability of science and, in the end, even behavioral science to make changes in time. He expressed a gathering despair about the future of the world.[53]

He had ruminated on that prospect since World War II. Project Pigeon had been an attempt to build a device to help end a horrific war. *Walden Two* had outlined a new way of life for postwar America, a new beginning via a powerful technology—behavioral science—that avoided the waste, gender inequities, and unrewarding work of an urban industrial society. By the early 1950s he expressed his social concern and the hopes for his science more directly: "Two exhausting wars in a single half century have given no assurance of lasting peace," he wrote. "Dreams of progress toward a higher civilization have been shattered by the spectacle of the murder of millions of innocent people. The worst may be still to come." Moreover, conventional technologies, however well meaning, often hindered solutions to social problems: "The application of science prevents famines and plagues, and lowers death rates—only to populate the earth beyond the reach of government control."[54] Industry developed among people unprepared for its cultural disruptions, while leaving unemployed others who had managed to adjust when economic strategies shifted.

On the other hand, the history of science showed a cumulative progress in physics, chemistry, and biology. Behavioral science had also progressed to the point where it could now be applied to human society to build a viable future. Skinner's optimism for using behavioral technology to avoid the catastrophes of the past and perhaps worse ones in the future peaked in the late 1950s and early 1960s. The massive success of the technology of teaching was crucial. But the failure of a more conventional technology—that is, American industry—to produce and market sufficient numbers of teaching machines, plus the resistance by the educational establishment, were major blows to his optimism as a social inventor.

By the late 1960s his fears for the future had become more urgent. *Beyond Freedom and Dignity* began by referring to "the terrifying problems that face the world today," problems that had become even more unresolvable because of the continuing successes of certain well-meaning but counterproductive technologies. "It is disheartening to find that technology is increasingly at fault. Sanitation and medicine have made the problems of population more acute, war has acquired a new horror with the invention of nuclear weapons, and the affluent pursuit of happiness is largely responsible for pollution."[55] Yet he still believed a "technology of behavior" could still make a difference: "We could solve our problems quickly enough if we could adjust the growth of the world's population as precisely as we adjust the course of a spaceship or improve agriculture and industry with some of the confidence with which we accelerate high-energy particles, or move toward a peaceful world with something like the steady progress with which physics has approached absolute zero."[56]

But the prospect of behavioral technology controlling population growth or defusing international conflict seemed to many either ridiculous or frightening rather than reassuring: "That is how far we are from 'understanding human issues' in the sense in which physics and biology understand their fields, and how far we are from preventing the catastrophe toward which the world seems to be inexorably moving." An understanding of human behavior still deferred to a humanistic tradition not much changed from that of the Greek philosophers, and although the physical and biological sciences had made immense strides, "our practices in government, education, and much of economics . . . have not greatly improved."[57]

Nonetheless, *Beyond Freedom and Dignity* was not written by a doomsayer. The very confidence with which Skinner insisted that behavioral technology could design a livable future underscored his scientific optimism. When his old friend Fred Keller wrote about his disappointment with many of the papers given for Division 25 at the APA national meeting in 1969 and his growing realization that he received "more real appreciation from physicists and engineers than I got anywhere from psychologists," Skinner did not share his pessimism.[58] "Your bitterness bothers me," he replied. "You and I have been in on one of the most exciting scientific developments in the twentieth century. Of course, it doesn't always go as we should like to have it, but what the hell? Truth will prevail—I am still idiot enough to believe that!"[59] The great question, however, was whether there would be enough time for truth to prevail.

A note of despair appeared in 1973. In a paper portentously entitled "Are We Free to Have a Future?" he continued to emphasize the critical necessity of immediately addressing global problems, especially overpopulation and ecological destruction. But he also noted: "The trend is certainly ominous, and Cassandra, who always prophesizes disaster, may again be right. If so, it will be for the last time. If she is right now, there will be no more prophecies of any kind."[60] Those cultural practices that had encouraged survival in the past had created a present less likely to produce new practices needed to sustain a future: "Our extraordinary commitment to immediate gratification has served the species well. The powerful reinforcing effects of . . . food, sexual contact, and . . . aggress[ion] . . . have had great survival value. Without them the species would probably not be here today, but under current conditions they are almost as nonfunctional as drugs, leading not to survival but to obesity and waste, to overpopulation and war."[61]

Still, he remained hopeful through the late 1970s—at least publicly. "I think you are right that the behaviorist position is at last beginning to be understood," he wrote a correspondent in 1978. And even though "newspaper editorials associate behaviorism with cattle prods . . . I am more optimistic than ever that we shall be able to save ourselves in time."[62] Only a few years later he was told that he seemed to be turning more pessimistic about the ability of humans to do what was necessary to ensure a future, and he readily agreed.[63]

Commenting in a new preface for *Beyond Freedom and Dignity* in 1988, he noted that when he wrote the book he thought we would be able to find "current surrogates" for the consequences of our acts that now threaten us:

> Give people reasons for having only a few children or none at all and remove the reasons why they have so many. Promote ways of life which are less consuming and less polluting. Reduce aggression and the likelihood of war by taking a smaller share of the wealth of the world. A science of behavior would spawn the technology needed to make changes of that sort, and I thought the science was aborning.[64]

Five months before his death, Skinner commented on the impending catastrophe—not only the end of the human species but "the end of the world." He was certain that the planet had already sealed its fate. This somber conclusion came after learning he was dying of leukemia, but he insisted his assessment had no relationship to his physical condition or to the usual end-of-century doomsday predictions. "I've always known I was going to die," he said with a slight smile. He seemed just as certain that the world would end:

> I think that the world is going to do what Shakespeare put in that line in his sonnet, "the world is going to be consumed by that which it was nourished by." I think evolution is random, accidental, no design in it at all. I think with the evolution of vocal musculature which made it possible for human beings to talk about the world, [to] have science and [culture], . . . that that didn't have built

> into it enough to take the future into account. And I am sure, quite sure, we have already passed the point of no return. . . . I can't imagine anything that will prevent the sheer destruction of the world as a planet long before it needs to be destroyed if there had evolved some mechanism which made it possible to take the future into account.

Although "it would take a long, long time," the reproduction of life would "eventually use up all the water and oxygen in the world." What hastened the process was the destructive power of the wrong kind of technology. "Inventing machines that exhaust [air and water] faster than people do it has accelerated it all. . . . That is the way it [the world] is going to stop."[65] Machine technology and human reproduction had been so overwhelmingly successful that they had overrun whatever opportunity behavioral technology might have had of altering or at least postponing the outcome.

IF SKINNER WAS CORRECT in predicting the rapidly approaching demise of the living planet and if one views behavioral technology as a missed opportunity to change that outcome, then his legacy is unspeakably tragic, transcending any legacy he may have perpetuated. Nonetheless, his connection with the culture that produced him is striking. He had grown to maturity in a progressive early-twentieth-century America that confidently predicted the end of war and the coming of a Christian brotherhood of humanity. But he had also witnessed two horrific international conflicts, a crushing national depression, and massive American cultural discontent and disorder. For most of his adult life Skinner believed—the last years notwithstanding—that "America may make its greatest contribution . . . by developing a better 'mechanism for living together' and the discipline most directly involved will be psychology as a science."[66] This was a faith for a better future based on prospects for designing a better culture—but it corresponded to a general, if anxious, American hopefulness, which assumed that national ingenuity, resourcefulness, and cooperation would ensure a better future.

Both soaring optimism and doomsday gloom had long been a part of the American tradition. New England Puritans had regularly forecasted both the millennium and the destruction of the world. A republican culture in the eighteenth century had entertained high hopes for a new Eden while harboring dark fears of the collapse of civilization if mankind's best hope, America, fell to the lure of self-indulgence. Religious sects in the nineteenth century provided new versions of the Puritan millennial-doomsday theme. But in all these traditions God was in control, and although human behavior certainly mattered, the ultimate future was God's to determine, not man's.[67] Skinner's final conviction that evolution did not build into the human species a mechanism strong enough to take the future into account appealed to biology as the originating determiner. God had nothing to do with the outcome; evolution was random natural selection. The power to design the future, even though biological evolution was accidentally determined, marked a central contradiction in Skinner's thinking. He, like most Americans for much of our history, preferred to emphasize the ability of people to shape their environments—except that he reversed the relationship: Environment determined how people acted rather than people freely creating their environments. Yet people still acted. Indeed, they needed to act in world-preserving ways.

Skinner explained in a clearly prophetic voice the relationship between choice and determinism to a sixteen-year-old correspondent who had just read *Beyond Freedom and Dignity:*

> We act in such ways because our environments have determined that we shall do so. It is always the environment which must be taken into account. A culture is an environment and if it induces people to behave in ways which strengthen the culture it will be more likely to solve its problems, meet emergencies and survive. . . . You and I are both strongly inclined to act with respect to the future of mankind because we have lived in an environment which has hit upon devices which strengthen such behavior. If our culture fails to induce others to do the same it is on its way out.[68]

Skinner did not actually deny the American heritage that insisted it was one's birthright to be able to shape one's future and yet also accept the tide of circumstance—whether God's will, destiny, or circumstance. Throughout his life Skinner readily acknowledged the role accident had played in determining the discovery of his science as well as the major turning points in his life, yet all the while maintaining that one should act—albeit because of the determining effect of environment—to design one's world. Skinner's legacy was not far from the old-fashioned American heritage; he wanted action and yet accepted fate. And, like most Americans, he remained for most of his life an optimist, a man who preferred to emphasize hopes and dreams for the future rather than predicaments, resignations, and dead ends.

Yet Skinner's legacy was novel and so challenging that to many it seemed a threat to the America they knew and loved. Skinner was operating in a different realm; he was using familiar words like *freedom* and *control* in unfamiliar ways. He often made such comments as: "man is not free in the sense that his behavior is undetermined. He has achieved a considerable freedom from the kinds of control which lead us to strike for freedom, and I am, of course, all in favor of that."[69] Or, denying that he wanted to impose controls, he maintained, "I am simply interested in improving those which now exist."[70] By concentrating on freedom and control in the context of behavior, rather than in the more traditional contexts of religion, government, or economics, he was able to redefine what Americans did and, more important, what Americans should do to have a viable culture. That was risky because it amounted to telling Americans *how* to behave.

Skinner, however, did not nationalize his science in the sense that it was meant only for Americans. Perhaps if he had tried to discuss his ideas in more familiar ways, with more deference to the traditions he was challenging, his social inventions might have met less resistance. But it did not matter in terms of a science of behavior whether the behaving individual was American, African, Latin, or Asian, or, for that matter, male or female. That was perplexing. How could he claim his behavioral science was humanistic when it seemed to discount these traditional distinctions? Yet what could be more universal, more democratic, more American, than promoting the unity in the diversity of humankind? What he offered was a "much nobler conception of man . . . through new methods of science."[71]

Moreover, it was the individual's behavior that Skinner wanted to reshape; he never really applied behavioral technology to groups in the sense that he wanted crowd control or any form of social revolution. He loathed mass solutions to social problems. Always his emphasis was on providing environments that reinforced and hence beneficially shaped individual behaviors. Groups or, more appropriately, cultures were transformed when individual behaviors were transformed; and that, too, was squarely in the American tradition of putting the individual's interest before society's—a tradition firmly established in the mid-nineteenth century.

On this point, however, Skinner's objective was not the preservation of the individual but the creation and perpetuation of a new kind of individual whose *behavior* could ensure cultural survival. Gone were the old standbys of the traditional individualism: character traits, conscience, and the inner man. Gone too were the traditional explanations of fate and accident, such as God and the unconscious mind. And soon, too, he hoped, would disappear the failing urban industrial environments—the faceless, aesthetically ugly, bureaucratically controlled cities, with their contrived reinforcers, such as wages. These would be replaced with smaller communities with life-enhancing and ecologically preserving technologies that promised an end to punitive environments, whether they be maintained by governments, economies, religions, or conventional technologies—a world, of course, much like that in *Walden Two.* "I am not trying to change people," Skinner insisted. "All I want to do is change the world in which they live."[72]

NOTES

The following abbreviations are used for frequently cited names and for archival collections.

APS	Arthur Percy Saunders
BA	Basement Archives at B. F. Skinner's home
BFS	B. F. Skinner
EGB	Edwin G. Boring
FSK	Fred S. Keller
HA	Harvard Archives, Harvard University
HCA	Hamilton College Library Archives
JWJ	James Weston Judd
RC	Radcliffe College, Yvonne Blue Skinner Collection, Arthur and Elizabeth Schlesinger Library on the History of Women in America
UM	University of Minnesota, R. M. Elliott Papers, Walters Library Archives
YBS	Yvonne (Eve) Blue Skinner (abbreviation not used when referring to Yvonne Blue before her marriage to B. F. Skinner)

CHAPTER 1

1. BFS, "A Winter's Day," Mar. 1, 1970, BA.
2. BFS, "The Autobiography," Mar. 28, 1974; and "For the Autobiography," Dec. 15, 1965, BA.
3. BFS to Mrs. C. B. Brownell, June 14, 1974, HA.
4. BFS, "The Importance of Disorder," Mar. 31, 1966, BA; and Interview with BFS, Dec. 19, 1988.
5. BFS, "Importance of Disorder."
6. BFS, "Jules Verne," June 15, 1971, BA.
7. "A Case History in Scientific Method," *American Psychologist* 11 (1956): 221–33.
8. George R. Taylor, *The Transportation Revolution: 1815–1860* (New York: Rinehart, 1951). Mrs. Arthur Webb, "History of Susquehanna, Pa." in *Susquehanna*

Cheers 100 Years, Susquehanna Centennial, Aug. 16–22, 1953. Other sources on Susquehanna include the following: Rhamanthus M. Stocker, *Centennial History of Susquehanna County, Pennsylvania* (Baltimore: Regional, 1974); *Triennial Atlas and Plat Book: Susquehanna, Pennsylvania* (Rockford, Ill.: Rockford Map Publishers, 1967); *SOLIDA* (Susquehanna, Oakland, Lanesboro Industrial Development Association), Susquehanna Free Library.

9. Sister M. St. Francis to BFS, Feb. 28, 1977, HA.
10. Interview with Roland Hendricks, Susquehanna, Pa., Aug. 16, 1990.
11. Robert N. Yetter to BFS, Jan. 28, 1977, HA. Yetter was the pastor of the Presbyterian Church in Susquehanna from 1956 to 1962.
12. Skinner family scrapbooks, 1903–1916 and 1916–1922, BA.
13. BFS, *Particulars of My Life* (Washington Square, N.Y.: New York University Press, 1979), p. 11.
14. Ibid., p. 6.
15. BFS, untitled note, Sept. 26, 1968, BA.
16. BFS, "The Autobiography," June 19, 1972, BA.
17. Virginia M. Packard to BFS, Sept. 8, 1979, HA.
18. BFS, "Historical Religious Note," [1926], BA; and *Particulars,* p. 60.
19. BFS, *Particulars,* p. 13.
20. BFS, "My Grandparents' House on Myrtle Street," Apr. 18, 1966, BA.
21. BFS, *Particulars,* pp. 15–16.
22. BFS, "Revision," Sept. 21, 1968; and "Healer," Oct. 24, 1965, BA.
23. BFS, *Particulars,* p. 14.
24. Ibid., p. 8.
25. Ibid., p. 12.
26. *Susquehanna Transcript,* May 1, 1902, BA.
27. Skinner family scrapbook, 1899–1900, BA.
28. BFS, *Particulars,* p. 21.
29. See Clyde Griffen, "The Progressive Ethos" in *The Development of an American Culture,* ed. Stanley Coben and Lorman Ratner (Englewood Cliffs, N.J.: Prentice-Hall, 1970), pp. 120–49.
30. Skinner family scrapbooks, 1903–1916 and 1916–1922, BA.
31. *Susquehanna Transcript,* Jan. 16, 1904, BA.
32. BFS, *Particulars,* p. 23.
33. BFS, "My Parents and Health," Apr. 18, 1966, BA.
34. BFS, "#26," Dec. 15, 1965, BA.
35. BFS, *A Matter of Consequences* (Washington Square, N.Y.: New York University Press, 1984), p. 20.
36. Sister M. St. Francis to BFS, May 29, 1975, HA.
37. BFS, "My Father's Aspirations," n.d. BA.
38. BFS, "New York," Mar. 24, 1971, BA.
39. BFS, *Particulars,* p. 44.
40. Ibid., pp. 45–46; and BFS, "Auto," Mar. 24, 1971, BA.
41. BFS, "Presentiment," July 23, 1967, BA.

42. B. F. Skinner: "A World of Difference," Transcript of PBS television special, n.d., HA.
43. Ibid.
44. BFS, "Early Aversive Control," Dec. 19, 1971; and "Note for the Autobiography," Jan. 23, 1966, BA.
45. BFS, "My Parents and Health."
46. Nathan Hale, Jr., *Freud and the Americans: The Beginnings of Psychoanalysis in the United States, 1876–1917* (New York: Oxford University Press, 1971).
47. BFS, "My Brother," Feb. 19, 1966, BA.
48. BFS, "Sketch for an Autobiography," unpublished manuscript [Early 1960s?], in the possession of Daniel Fallon, Texas A&M University, College Station, Tex. A revised version of this draft appeared as "B. F. Skinner (An Autobiography)," in *A History of Psychology in Autobiography,* vol. 5, ed. E. G. Boring and G. Lindzey (New York: Appleton-Century-Crofts, 1967), pp. 387–413.
49. BFS, "Doing Good," Jan. 30, 1968, BA.
50. BFS, "Filial Disloyalty?" Feb. 19, 1966, BA.
51. BFS, "Sentimentality," Apr. 18, 1966, BA.
52. "A World of Difference."
53. Sister M. St. Francis to BFS, May 29, 1975, HA.
54. BFS, untitled note, n.d., probably written in early 1970s, BA.
55. BFS, "Adagietto," Feb. 14, 1970, BA.
56. BFS, "My Brother."
57. BFS, "Echoic Teasing," May 9, 1968, BA.
58. BFS, "My Brother."
59. BFS, "The Inquiring Mind" [ca. 1970], BA.
60. Sister M. St. Francis to BFS, May 29, 1975, HA.
61. "B. F. Skinner . . . an Autobiography," in *Festschrift for B. F. Skinner,* ed. P. W. Dews, (New York: Irvington, 1970), p. 2.
62. Jeanette Brownell to BFS, Apr. 19, 1976, HA.
63. BFS, untitled, Nov. 6, 1971; BFS, "Fifty-year-old memories." Apr. 7, 1969, BA.
64. BFS, "Fifty-year-old memories," Apr. 7, 1969, BA.
65. BFS, "Mechanical Aptitudes," Mar. 30, 1966, BA.
66. BFS, "Autobiography," Dec. 21, 1971, BA.
67. Otis H. Chidester to BFS [ca. 1975], HA.
68. BFS, "Levitation," June 18, 1971, BA.
69. BFS to John Poluhowich, Nov. 30, 1977, HA.
70. Beatrice Seary Autes to BFS, n.d., BA.
71. BFS, "Animal Lore," Mar. 31, 1966, BA.
72. BFS, "Additional Nature Lore," Oct. 12, 1968, BA.
73. BFS, "Sketch for an Autobiography."
74. BFS, "The Little Books," Mar. 16, 1966, BA.
75. BFS, untitled, Aug. 31, 1969, BA.
76. Ibid.
77. BFS to Jim Read, Jan. 28, 1959, HA.

78. BFS, "A Box to Hide In," Mar. 16, 1966, BA. This note had the same title as an earlier one, dated Mar. 8, 1962, that Skinner wanted to amend, but he noted, "I have not seen it lately."
79. BFS, "Autobiography. Music," Oct. 30, 1971, BA.
80. BFS, "Backward O Backward," Feb. 7, 1966, BA.
81. BFS, "Ward Palmer," Apr. 1, 1966, BA; and Sister M. St. Francis to BFS, Nov. 14, 1974, HA.
82. Jeanette Brownell to BFS, Apr. 19, 1976, HA.
83. BFS, "Early Years," Aug. 23, 1970, BA.
84. BFS, "Sketch for an Autobiography."
85. BFS, "Miss Keefe," Feb. 23, 1967, BA.
86. BFS, "Early Years."
87. BFS's high school transcripts are located in the Hamilton College Library Archives, Clinton, N.Y.
88. BFS, "Early Years."
89. BFS to Rev. Robert N. Yetter, Feb. 15, 1977, HA.
90. Mary Graves's notebook, n.d., BA.
91. BFS, "Sketch for an Autobiography."
92. BFS to Arthur G. Lawson, Oct. 9, 1979, HA.
93. See Carl Van Doren, "The Revolt from the Village: 1920," *Nation,* Oct. 12, 1921, pp. 307–412, and Anthony Channell Hilfer, *The Revolt from the Village, 1915–1930* (Chapel Hill: University of North Carolina Press, 1969).
94. BFS, "Community," Mar. 26, 1968, BA.
95. BFS, "Need for Approval," Dec. 15, 1967, BA.
96. Sister M. St. Francis to BFS, Nov. 14, 1974, and Nov. 5, 1974, HA.
97. BFS, "Autobiography," Mar. 28, 1970, BA.
98. BFS, "My Diary," May 27, 1921, BA.
99. Otis H. Chidester to BFS [ca. 1975], HA.
100. BFS, "Note for the Autobiography," Jan. 23, 1966, BA.
101. BFS, *Particulars,* p. 183. See also Arthur S. Link, "What Happened to the Progressive Movement in the 1920s?" *American Historical Review* 64 (July 1959): 833–51; and Loren Baritz, "The Culture of the Twenties," in Coben and Ratner, *The Development of an American Culture,* pp. 150–78.
102. Skinner family scrapbook, 1916–1922, BA.
103. BFS, *Particulars,* p. 184; and BFS, "Autobiography," Feb. 21, 1974, BA.
104. John K. Hutchens, "Way Back Then, and Later," *Hamilton Alumni Review,* 42 (Dec. 1976): 12–18, 13, HCA.
105. Certificate of Preparation for Hamilton College, Aug. 17, 1922, HA.
106. John N. Ogelthorpe to Frederick C. Ferry, Sept. 20, 1922, BA.
107. BFS, "My Attitude Toward My Parents," Feb. 19, 1969, BA (BFS's emphasis).
108. BFS to Jeanette Brownell, Sept. 9, 1976, HA.
109. BFS, "Caste-System," Jan. 30, 1977, BA.
110. BFS to Ted Odell, Jan. 27, 1970, HA.
111. "A World of Difference."

CHAPTER 2

1. See Whitney R. Cross's classic, *The Burned-Over District: The Social and Intellectual History of Enthusiastic Religion in Western New York, 1800–1850* (New York: Harper & Row, 1965).
2. "B. F. Skinner: A World of Difference," Transcript of PBS television special, n.d., HA.
3. For details on the transition from academy to college, see Walter Pilkington, *Hamilton College, 1812–1962* (Deposit, N.Y.: Courier Printing, 1962), pp. 1–97.
4. John K. Hutchens, "Way Back Then, and Later," *Hamilton Alumni Review,* 42 (Dec. 1976): 12–18, 12.
5. BFS to his parents, Sept. 19, 1922, BA.
6. Gramp to Stewart, May 14, 1974, BA.
7. Hutchens, "Way Back Then," p. 12.
8. Pilkington, *Hamilton College,* pp. 229–40.
9. Hutchens, "Way Back Then," p. 16.
10. Hamilton College transcripts, BA.
11. Interview with BFS, Aug. 13, 1990.
12. Hutchens, "Way Back Then," p. 16.
13. BFS, *Particulars of My Life* (Washington Square, N.Y.: New York University Press, 1979), p. 198.
14. BFS, "Social Stimulus and Control," Aug. 26, 1964, BA.
15. For more on the excessive hazing at Hamilton, see Pilkington, *Hamilton College,* 231–32. The quotations in the text are from an interview with Peter B. Daymont, Aug. 6, 1990; and BFS to Silvia Saunders, Feb. 8, 1977, HA.
16. BFS, "Self-confidence," Sept. 7, 1963, BA.
17. BFS, *Particulars,* p. 193.
18. BFS, "Caste-System," Jan. 30, 1977, BA.
19. Hutchens, "Way Back Then," p. 13.
20. Gramp to Stewart. Fred tried to convince his philosophy professor "that [Fred] is the center of the universe." The professor responded by "telling him that he is an egotist!" *Hamiltonian,* 1926, p. 50, HCA.
21. Interview with BFS, Mar. 7, 1990.
22. Hamilton College transcripts, BA.
23. BFS, *Particulars,* pp. 206–7.
24. BFS, "A Change of View After One Year at College," English composition theme [Spring 1923], BA.
25. Ibid.
26. *Susquehanna Transcript,* Apr. 9, 1923, BA.
27. BFS, "My Father and My Brother," Feb. 19, 1966, BA.
28. BFS, *Particulars,* p. 209.
29. BFS, "My Father and My Brother."
30. BFS, *Particulars,* p. 210.
31. Interview with BFS, Mar. 7, 1990.

32. BFS, *Particulars,* p. 210.
33. BFS, "Ebbie's Death," Nov. 6, 1967, BA.
34. BFS, "My Attitude Toward My Parents," Feb. 19, 1969, BA.
35. BFS, *Particulars,* p. 212.
36. Skinner family scrapbook, 1923–1930, BA.
37. BFS, *Particulars,* p. 213.
38. BFS, "Upper Class," Jan. 12, 1967, BA.
39. BFS, *Particulars,* p. 214.
40. *New York Times,* Aug. 15, 1953, HCA.
41. Arthur Percy Saunders's diaries 1877–1944, HCA.
42. Frank K. Lorenz to Cathy Abernathy, May 24, 1977, HCA.
43. BFS to Margaret Briggs, Jan. 24, 1967, HA; and Interview with Olivia Saunders Wood, Aug. 6, 1990.
44. Interview with BFS, Mar. 7, 1990. The scandal was also fictionalized in Paul Koonradt's *Dance Out the Answer* (New York: Longmans, Green, 1932).
45. Arthur Percy Saunders's diary, Apr. 29, 1926, HCA.
46. BFS, "A Better World," Dec. 6, 1984, BA.
47. For a discussion of American business values, see Peter Baida, *Poor Richard's Legacy: American Business Values from Benjamin Franklin to Donald Trump* (New York: Morrow, 1990).
48. See F. O. Matthiessen, *American Renaissance: Art and Expression in the Age of Emerson and Whitman* (New York: Oxford University Press, 1941).
49. BFS, *Particulars,* p. 218.
50. *Hamiltonian,* 1926, p. 50.
51. Interview with BFS, Aug. 13, 1990.
52. BFS, *Particulars,* p. 218.
53. Interview with Olivia Saunders Wood, Aug. 6, 1990.
54. Interview with BFS, Mar. 5, 1990.
55. Hamilton College transcripts, BA.
56. Interview with BFS, Aug. 13, 1990.
57. *Hamiltonian,* 1926, p. 41.
58. Ibid.
59. Hutchens, "Way Back Then," p. 13.
60. Hamilton College transcripts, BA.
61. BFS, *Particulars,* p. 295.
62. Ibid., pp. 230–31.
63. John K. Hutchens to BFS, Aug. 4 and July 14, 1974, HA.
64. Hamilton College transcripts, BA.
65. *Royal Gaboon,* Oct. 1925, pp. 5, 14, HCA.
66. BFS, *Particulars,* p. 254.
67. BFS, "Making a Break," May 2, 1969, BA. Skinner quotes generously from "Elsa" in *Particulars,* pp. 252–53.
68. Interview with BFS, Dec. 11, 1989.

69. BFS to Otis H. Chidester, Jan. 24, 1975; and Otis H. Chidester to BFS, [no date, but in response to BFS's letter], HA.
70. BFS, *Particulars,* p. 237.
71. Ibid., pp. 245–46.
72. Ibid., p. 246.
73. Ibid., p. 247.
74. Ibid., p. 248.
75. Ibid., pp. 248–49. The letter is also published in *Selected Letters of Robert Frost,* ed. Lawrence Thompson (New York: Holt, Rinehart & Winston, 1964). The story Frost liked best was called "The Laugh," a tale of miscommunication between husband and wife.
76. Interview with BFS, Mar. 5, 1990.
77. BFS, *Particulars,* p. 255.
78. Translation by Richard A. LaFleur, July 1989, BA.
79. Interview with BFS, Aug. 13, 1990.

CHAPTER 3

1. For Skinner's reconstruction of this period, see *Particulars of My Life* (Washington Square, N.Y.: New York University Press, 1979), pp. 262–303.
2. Ibid., p. 262. For a comprehensive discussion of the little magazines, see Frederic J. Allen, Charles Allen, and Carolyn F. Ulrich, *The Little Magazines: A History and a Bibliography* (Princeton, N.J.: Princeton University Press, 1946).
3. BFS to APS, July 5, 1926, HCA.
4. See Arthur Machen, *A Hill of Dreams* (London: Grant Richards, 1907; reprint, London: Baker, 1968). For Skinner's recollection of the book, see BFS, "I Love My Work?" Aug. 13, 1986, BA. BFS, "Sketch for an Autobiography," unpublished manuscript [early 1960s?], in the possession of Daniel Fallon, Texas A&M University, College Station, Tex., p. 10.
5. BFS, "On Writing an Autobiography," n.d., HA. See also S. R. Coleman, "B. F. Skinner, 1926–1928: From Literature to Psychology," *Behavior Analyst* 8 (Spring 1985): 77–85. Coleman recognized that Skinner's "reading and conversation might be only *consequences* or reflections of processes . . . which lie outside the strictly literary" [Coleman's emphasis] (p. 77). Another writer ignored the literary influence in favor of understanding the Dark Year as the beginning of a protracted identity crisis that was not resolved until Skinner wrote *Walden Two;* hence, he was more interested in signs of a career crisis than in reasons that Skinner became a behaviorist. See Alan Elms, "Skinner's Dark Year and *Walden Two,*" *American Psychologist* 36 (1981): 470–79. Also useful is a thoughtful analysis emphasizing the social dimension of Skinner's thinking during the Dark Year. See Nils Wiklander, *From Laboratory to Utopia: An Inquiry in the Early Psychology and Social Philosophy of B. F. Skinner* (Göteborg, Sweden: Göteborg University Press, Arachne Series, 1989).

6. BFS, "Romanticism," Jan. 22, 1983, BA. Skinner remembered hearing about Murray's remark when he was a graduate student.
7. See David A. Hollinger, "The Knower and the Artificer," in *Modernist Culture and American Life,* ed. Daniel Joseph Singal (Belmont, Calif.: Wadsworth, 1991), pp. 42–69.
8. BFS to APS, Aug. 16, 1926, HCA.
9. Ibid.
10. "Sketch for an Autobiography," p. 10. Actually, Charles Dickens, in the preface to *The Old Curiosity Shop,* had originally made this observation about the character Mrs. Mackenzie, in Thackeray's *The Newcomes.* See BFS, *Particulars,* p. 291. Skinner's Susquehanna friend was Annette Kane, who later became a nun and wrote an unpublished utopian novel.
11. BFS, *Particulars,* p. 267.
12. BFS to APS, Aug. 16, 1926, HCA.
13. BFS to APS, Nov. 14, 1926, HCA.
14. BFS to APS, Aug. 16, 1926, HCA.
15. Ibid.
16. Ibid.
17. BFS, "Intellectual Suicide," Nov. 11, 1962, BA. "This was a favorite theme of mine in the late 20's—under the influence of my first exposure to Behaviorism."
18. See C. K. Ogden and I. A. Richards, *The Meaning of Meaning* (New York: Harcourt, Brace, 1923); and E. A. Burtt, *The Metaphysical Foundations of Modern Physical Science: A Historical and Critical Essay* (London: Paul, Trench, & Trubner, 1925). Skinner borrowed the Burtt book from the Hamilton College Library and took notes on it. He believed it stimulated "serious thinking," and he held that his "assertions must be understood to come from no authority but [his] own feeling" (*Particulars,* pp. 280–81). Of course, that was precisely the "authority" that behavioral science was challenging.
19. Joseph Wood Krutch, *The Modern Temper: A Study and a Confession* (New York: Harcourt Brace, 1929). For an analysis of Krutch and the tension he perceives between artist and scientist, see Hollinger, "Knower and Artificer," pp. 54–55.
20. Joseph W. Krutch, *The Measure of Man: On Freedom, Human Values, Survival, and the Modern Temper* (Indianapolis: Bobbs-Merrill, 1954).
21. For Russell's attitude toward science in its historical context, see Frederic J. Hoffman, *The Twenties: American Writing in the Postwar Decade* (New York: Free Press, 1965), pp. 277–81.
22. Daniel Joseph Singal has recently discussed the difficulty of interpreting American modernism as simply antibourgeois, given a strong progressive, pro-science strain within the parameters of that historical setting. See Singal, "Towards a Definition of American Modernism," in Singal, ed., *Modernist Culture,* pp. 1–27. Also helpful are Loren Baritz, "The Culture of the Twenties" in *The Development of an American Culture,* ed. Stanley Coben and Lorman Ratner

(New York: St. Martin's Press, 1983); and Malcolm Cowley, *Exile's Return* (New York: Viking, 1951).

23. BFS, "Books That Have Influenced Me," n.d., BA.
24. Interview with BFS, Dec. 14, 1989.
25. The possibility of Skinner acting as a scientist before he became one is suggestive of Erik Erikson's generalization in his study of Martin Luther: namely, that the crisis in "a young man's life may be reached when he half-realizes that he is fatally over-committed to what he is not" (*Young Man Luther: A Study in Psychoanalysis and History* [New York: Norton, 1958], p. 43). If Skinner was in fact a romantic who defended himself with science, the discovery that he *was* a scientist was a pivotal moment.
26. BFS, "Self-confidence," Sept. 7, 1963, BA.
27. BFS, "Why I Am Deserting Writing for Several Years," quoted in *Particulars,* pp. 264–65.
28. BFS, "Self-confidence." Sept. 7, 1963, BA.
29. BFS to APS, Sept. 29, 1926, HCA.
30. BFS to APS, Jan. 9, 1927, HCA.
31. BFS to APS, Dec. 15, 1926, HCA.
32. BFS, *Particulars,* p. 264.
33. BFS to APS, Feb. 22, 1927, HCA.
34. Fred was sarcastically critical of an unfavorable review of *Clissold* in the *Saturday Review of Literature,* describing the review as "excellent slop," while applauding Wells for "so nicely avoid[ing] formulating, [and] concluding." BFS to APS, Sept. 29, 1926, HA.
35. BFS to APS, Dec. 15, 1926, HCA.
36. Ibid.
37. Ibid.
38. For more on Loeb, see Philip J. Pauly's fine *Controlling Life: Jacques Loeb and the Engineering Ideal in Biology* (Berkeley: University of California Press, 1990.) Mach's great influence on Skinner has been skillfully analyzed by Laurence D. Smith in *Behaviorism and Logical Positivism: A Reassessment of the Alliance* (Stanford, Calif.: Stanford University Press, 1986), pp. 258–97.
39. Interview with BFS, Mar. 9, 1990.
40. See Pauly, *Controlling Life,* pp. 188–89, 196; and BFS, *The Shaping of a Behaviorist* (New York: Knopf, 1979), pp. 16–17.
41. See Smith, *Behaviorism and Logical Positivism,* pp. 271–72.
42. BFS, *Particulars,* p. 259.
43. Interview with BFS, Aug. 13, 1990.
44. BFS, untitled, Dec. 19, 1981, BA.
45. BFS, *Particulars,* p. 278.
46. Interview with BFS, Aug. 13, 1990.
47. BFS, *Particulars,* p. 278.
48. BFS, "The Audience," Aug. 8, 1965, BA. His desire to be seen by his father as a "solid achiever" may have helped produce the duplicity.

49. BFS to APS, Sept. 29, 1926, HCA.
50. BFS to APS, Jan. 9, 1927, HCA.
51. Interview with BFS, Dec. 14, 1989.
52. BFS to APS, Jan. 24, 1927, HCA.
53. Ibid.
54. Indeed, Fred S. Keller, Skinner's longtime friend and fellow behaviorist, commented that during their graduate student days at Harvard, Skinner had read about behaviorism as a philosophy of science but had "said this [behaviorism] is not science" (Interview with FSK, Apr. 26, 1990, Chapel Hill, N.C.).
55. BFS to APS, Jan. 24, 1927, HCA.
56. BFS, *The Shaping of a Behaviorist,* p. 40.
57. BFS to APS, Feb. 22, 1927, HCA.
58. "Sketch for an Autobiography," p. 10.
59. BFS to APS, May 8, 1927, HCA.
60. Ibid.
61. See Donald Meyer, *The Positive Thinkers: Religion as Pop Psychology from Mary Baker Eddy to Oral Roberts* (New York: Pantheon, 1980), pp. 318–20.
62. See John W. Aldridge, "Afterthoughts on the 20's," *Commentary,* Nov. 1973, p. 40. Skinner never participated in the dissipations so fully, but he, perhaps more than any other American intellectual, would eventually be associated with the notion of "control" and the denial of "freedom."
63. BFS, "A Case History in Scientific Method," *American Psychologist* 11 (1956): 221–33.
64. BFS to APS, Apr. 8, 1927, HA; and Interview with BFS, Dec. 14, 1989.
65. BFS to APS, Nov. 2, 1927, HA.
66. BFS, *Particulars,* p. 302.
67. Interview with BFS, Mar. 7, 1990.
68. Caroline F. Ware, *Greenwich Village, 1920–1930* (New York: Harper & Row, 1935), pp. 235–63.
69. Interview with BFS, Mar. 7, 1990.
70. BFS, "Rhapsody in Blue," Mar. 4, 1955, BA.
71. Interview with BFS, Mar. 7, 1990; and BFS, *Particulars,* p. 310.
72. BFS to APS, May 28, 1928, HA.
73. For a sampling of the popularity of Freud among American intellectuals in the 1920s, see Hoffman, *The Twenties: American Writing,* pp. 203–36. For a study of the New World reception to Freud, see Nathan Hale, Jr., *Freud and the Americans: The Beginnings of Psychoanalysis in the United States, 1876–1917* (New York: Oxford University Press, 1971).
74. BFS to APS, May 28, 1928, HCA.
75. BFS, *Particulars,* p. 313.
76. Hugh Kenner makes a metaphoric connection in the history of modernity by linking American technological flight and European artistic freedom, specifically, the 1904 flight of the Wright brothers at Kitty Hawk, N.C., with James

Joyce's *A Portrait of the Artist as a Young Man* (*A Homemade World: The American Modernist Writers* [New York: Knopf, 1975], p. xii).

77. BFS, *Particulars,* pp. 317–18.
78. Ibid., p. 319.

CHAPTER 4

1. Robert W. White to BFS, Sept. 26, 1977, HA.
2. BFS to his parents [early fall 1928], BA.
3. BFS to APS, Sept. 26, 1928, HA.
4. BFS, Harvard notebook for 1928, HA.
5. BFS to his parents [early fall 1928], BA.
6. BFS to his parents [fall 1928], BA.
7. *Harvard University Gazette,* Mar. 17, 1956, p. 194.
8. See Philip J. Pauly, *Controlling Life: Jacques Loeb and the Engineering Ideal in Biology* (Berkeley: University of California Press, 1990).
9. BFS to his parents, "Friday" [late fall 1928], BA.
10. For an interesting discussion of Darwin, Romanes, and Morgan, see Robert A. Boakes, *From Darwin to Behaviorism: Psychology and the Minds of Animals* (Cambridge: Cambridge University Press, 1984), pp. 1–52.
11. A good discussion of developments in the young science of psychology in the late nineteenth and early twentieth centuries is Thomas Hardy Leahey's *History of Psychology: Main Currents in Psychological Thought* (Englewood Cliffs, N.J.: Prentice-Hall, 1987), pp. 138–299. See also Daniel W. Bjork, *The Compromised Scientist: William James in the Development of American Psychology* (New York: Columbia University Press, 1983).
12. John B. Watson, "Psychology as the Behaviorist Views It," *Psychological Review* 20 (1913): 158. For a recent biography of Watson, see Kerry W. Buckley's *Mechanical Man: John Broadus Watson and the Beginnings of Behaviorism* (New York: Guilford Press, 1989).
13. Interview with BFS, July 13, 1989. Skinner also recalled, "Although as a psychologist I was concerned with behavior that did not of necessity make me a behaviorist" (BFS, "Reminiscences of JEAB," *Journal of the Experimental Analysis of Behavior* 48 [Nov. 1987]: 448).
14. BFS, note [early 1930s], HA.
15. BFS, *The Shaping of a Behaviorist* (New York: Knopf, 1979), pp. 31–32.
16. Ibid.
17. James Dinsmoor to Daniel W. Bjork, June 12, 1992.
18. BFS to his parents, Jan. 1929, BA.
19. BFS to APS, Dec. 5 [1928], HA.
20. BFS to his parents [Jan. or Feb. 1929], BA.
21. BFS to his parents [early fall 1928], BA.
22. BFS to his parents [winter 1928–1929], BA.
23. BFS, *Shaping of a Behaviorist,* p. 31.

24. BFS, "Early Behaviorism," Feb. 26, 1972, BA.
25. Boakes, *From Darwin to Behaviorism,* pp. 58–68, 156–58.
26. Transcript of B. F. Skinner film sound track [mid- or late 1970s?], HA. Fred S. Keller also participated in the interviewing, and Keller called Hunter "a breath of fresh air."
27. BFS to Walter S. Hunter, Apr. 13, 1949, HA.
28. Boakes, *From Darwin to Behaviorism,* pp. 184–96.
29. BFS, *Shaping of a Behaviorist,* p. 31.
30. Ibid., p. 35.
31. Ibid.
32. Interview with FSK, Apr. 26, 1990, Chapel Hill, N.C.
33. Ibid.
34. Transcript of Skinner film sound track.
35. BFS, "Self-confidence," Sept. 7, 1963, BA.
36. Interview with BFS, Mar. 9, 1990.
37. BFS, *Shaping of a Behaviorist,* p. 3.
38. Ibid., p. 5.
39. Interview with FSK, Apr. 26, 1990.
40. BFS to William Skinner [Dec. 1928], BA.
41. BFS to his parents, Jan. 1929, BA.
42. BFS to his parents [fall 1928], BA.
43. BFS to Daniel N. Weiner, June 1, 1976, HA.
44. *Harvard University Gazette,* Mar. 17, 1956, p. 194.
45. Transcript of Skinner film sound track.
46. See a history of rat and maze psychology in Boakes, *From Darwin to Behaviorism,* pp. 143–48.
47. BFS, *Shaping of a Behaviorist,* p. 36.
48. Ibid., p. 36.
49. Ibid., p. 37. The kymograph was related to the spygmograph (1860) and the myrograph (1868). Invented by the Frenchman Étienne-Jules Marey, these instruments utilized the smoked paper drum to record, respectively, the human pulse-beat and the movements of a muscle. Wilhem Wundt and Hermann Helmholtz, two pioneers in late-nineteenth-century physiological psychology, found the myrograph essential to their efforts to quantify nerve and muscle movement. Although he would emphasize behavior rather than physiology, Skinner depended on these early efforts to make a visual record of movement (Siegried Giedion, *Mechanization Takes Command* [New York: Norton, 1969], pp. 17–30).
50. Interview with BFS, Mar. 9, 1990.
51. See Pauly, *Controlling Life,* pp. 41–47, 195–96; and Laurence D. Smith, "On Prediction and Control: B. F. Skinner and the Technological Ideal of Science," *American Psychologist* 47 (Feb. 1992): 216–23.
52. BFS, *Shaping of a Behaviorist,* p. 38.
53. Interview with FSK, Apr. 26, 1990.

54. BFS, *Shaping of a Behaviorist,* pp. 32–33.
55. Ibid., pp. 51–52.
56. Ibid., p. 53.
57. Ibid., p. 55.
58. BFS to his parents [1930], BA.
59. BFS, "A Case History in Scientific Method," *American Psychologist* 11 (1956): 221–33.
60. Interview with BFS, Dec. 15, 1989.
61. BFS, *Shaping of a Behaviorist,* p. 59.
62. BFS to his parents [late March 1930], BA.
63. Interview with BFS, Mar. 9, 1990.
64. Interview with BFS, Mar. 8, 1990.
65. Interview with BFS, Mar. 9, 1990.
66. Ibid.
67. BFS, "Historical Note," Apr. 24, 1966, BA.
68. Interview with FSK, Apr. 26, 1990.
69. BFS to his parents [Jan. 1930], BA.
70. BFS to his parents [1930], BA.
71. Transcript of Skinner film sound track.
72. BFS to his parents [fall 1929], BA.
73. BFS to Barbara Ross, Aug. 10, 1970, HA.
74. Edwin G. Boring to Dean [Paul] Buck, Jan. 15, 1951, HA.
75. BFS to his parents [fall 1929], BA.
76. BFS to his parents [spring 1930], BA.
77. One graduate student at Harvard recalled that Skinner also used a timer. See Richard Herrnstein, "Reminiscences Already," *Journal of the Experimental Analysis of Behavior* 48 (Nov. 1987): 448–53, esp. p. 450.
78. David Shakow, untitled note, n.d., HA.
79. Interview with FSK, Apr. 26, 1990.
80. For a detailed discussion of both the academic politics involved in approval of the dissertation and the intellectual issues that divided Boring and Skinner, see S. R. Coleman, "When Historians Disagree: B. F. Skinner and E. G. Boring, 1930," *Psychological Record* 35 (1985): 301–14.
81. The article, which was virtually identical to the theoretical part of his dissertation, appeared as "The Concept of the Reflex in the Description of Behavior" (*Journal of General Psychology* 5 [1931]: 427–548). Crozier, the editor of the *Journal of General Psychology,* encouraged Skinner to submit the article and recommended quick publication.
82. BFS, "The Concept of the Reflex," p. 455.
83. EGB to BFS, Oct. 13, 1930, HA.
84. BFS, "The Concept of the Reflex," p. 443.
85. Ibid., p. 455.
86. EGB to BFS, Oct. 13, 1930, HA.
87. BFS to David Krantz, Oct. 28, 1970, HA.

88. EGB to BFS, Oct. 13, 1930, HA.
89. Ibid.
90. Ibid.
91. BFS to EGB, Dec. 14, 1930. Quoted in Coleman, "When Historians Disagree," p. 311.
92. Interview with BFS, July 13, 1989.
93. EGB to BFS, Dec. 3, 1930.
94. Interview with BFS, Dec. 14, 1989. His reading prompted him to write a review (coauthored with William Crozier) of Franklin Fearing's *Reflex Action: A Study in the History of Physiological Psychology* (*Journal of General Psychology* 5 [1931]: 125–29).
95. Interview with BFS, Dec. 14, 1989.
96. BFS to James J. Gibson, July 6, 1977, HA.
97. BFS, *Upon Further Reflection* (Englewood Cliffs, N.J.: Prentice-Hall, 1987), p. 189.
98. BFS to C. G. Costello, Aug. 20, 1971, HA.
99. Percy Bridgman, *The Logic of Modern Physics* (New York: Macmillan, 1927).
100. Interview with BFS, Dec. 12, 1989.
101. Ibid.
102. For an excellent analysis of the intellectual connection between Mach and Skinner, see Laurence D. Smith, *Behaviorism and Logical Positivism: A Reassessment of the Alliance* (Stanford, Calif.: Stanford University Press, 1986), pp. 264–75.
103. Interview with BFS, Dec. 12, 1989. For more on Loeb and Crozier, see Pauly's *Controlling Life,* pp. 161, 183–85.
104. Transcript of Skinner film sound track.
105. BFS to APS, Jan. 6 [1931], HA.
106. William J. Robbins to BFS, Feb. 8, 1932, HA.
107. EGB to Frank A. Lillie, Dec. 19, 1930, HA.
108. *Boston Herald,* Apr. 12, 1933, HA; and Skinner family scrapbook 1934–1940, BA.
109. BFS to his parents, n.d., BA.
110. BFS, "Relevance," Feb. 10, 1970, BA.
111. BFS, "Being There at the Right Time," May 12, 1969, BA.
112. BFS, "Crozier" [n.d.], BA.

CHAPTER 5

1. Interviews with BFS, Mar. 9, 1990, and Aug. 15, 1990.
2. For a discussion of Tolman and Hull, as well as of Skinner, see Laurence D. Smith, *Behaviorism and Logical Positivism: A Reassessment of the Alliance* (Berkeley: University of California Press, 1986), pp. 301–17. Also helpful is James A. Dinsmoor, "In the Beginning . . . ," *Journal of the Experimental Analysis of Behavior* 50 (Sept. 1988): 287–96, esp. 293–94.

3. BFS to David Quartermain, Mar. 13, 1956, HA.
4. BFS, "The Generic Nature of the Concepts of Stimulus and Response," *Journal of General Psychology* 12 (1935): 40–65.
5. For a discussion of how Skinner solved the problem of variability after 1931, see S. R. Coleman, "Historical Context and Systematic Functions of the Concept of the Operant," *Behaviorism* 9 (Fall 1981): 207–26.
6. A stimulus, indeed, might still be there, but "none was operative at the time the behavior was observed" (BFS, "Two Types of Conditioned Reflex: A Reply to Konorski and Miller," *Journal of General Psychology* 16 [1937]: 274, 277–79).
7. See Robert Epstein, "A Listing of the Published Works of B. F. Skinner, with Notes and Comments," *Behaviorism* 5 (1977): 99–110, esp. 103–4. See also BFS, "Drive and Reflex Strength," *Journal of General Psychology* 6 (1932): 22–48; "On the Rate of Formation of a Conditioned Reflex," *Journal of General Psychology* 8 (1932): 274–86; "On the Rate of Extinction of a Conditioned Reflex," *Journal of General Psychology* 9 (1933): 114–29; "The Rate of Establishment of a Discrimination," *Journal of General Psychology* 9 (1933): 420–29.
8. FSK, *Summers and Sabbaticals: Selected Papers on Psychology and Education* (Champaign, Ill.: Research Press, 1977), p. 11.
9. BFS, *The Behavior of Organisms: An Experimental Analysis* (New York: Appleton-Century-Crofts, 1938).
10. For the history of the reflex, see R. M. Young, *Mind, Brain and Adaptation in the Nineteenth Century* (New York: Clarendon Press, 1970); and Coleman, "Historical Context and Systematic Functions," 208–11.
11. See J. Konorski and S. Miller, "On Two Types of Conditioned Reflex," *Journal of General Psychology* 12 [1935]: 66–77). In his rebuttal, Skinner defended reinforcement theory and first used the term *operant* ("Two Types of Conditioned Reflex: A Reply to Konorski and Miller," *Journal of General Psychology* 16 [1937]: 272–79). For a detailed discussion of the exchange, see Coleman, "Historical Context and Systematic Functions," 218–22.
12. For Skinner's lengthy discussion of the role of stimulus in both respondent and operant psychology, see *The Behavior of Organisms,* pp. 167–307.
13. Travis Thompson, "Retrospective Review: Benedictus Behavioral Analysis: B. F. Skinner's Magnum Opus at Fifty," *Contemporary Psychology* 33 (1988): 397–402.
14. Ernest R. Hilgard, "Review of *The Behavior of Organisms,*" *Psychological Bulletin* 36 (1939): 121–25.
15. BFS to David Krantz, May 16, 1970, HA.
16. BFS to David Krantz, Dec. 14, 1971, HA.
17. BFS to David Krantz, Oct. 28, 1970, HA.
18. Edward C. Tolman to BFS, Nov. 14, 1938, HA.
19. Walter S. Hunter to BFS, Nov. 28, 1938, HA.
20. Transcript of B. F. Skinner film sound track, n.d., HA. (FSK was also present.)
21. *Behavior of Organisms,* pp. 441–42.
22. Thompson, "Retrospective Review," p. 399.

23. BFS to FSK, Apr. [?], 1936, quoted in *The Shaping of a Behaviorist* (New York: Knopf, 1979), p. 183.
24. Richard M. Elliott to EGB, Feb. 2, 1931, R. M. Elliot Papers, Walters Library Archives, UM.
25. BFS to Eliot Hearst, July 13, 1976, HA. The "famous paper" in which Tolman used Skinner's "third variable" as an "intervening variable" was "Psychology versus Immediate Experience" (*Philosophy of Science* 2 [1935]: 356–80). Skinner remembered that "that may have been the point at which the experimental analysis of behavior parted company from what would be cognitive psychology" (BFS, *Recent Issues in the Analysis of Behavior* [Columbus, Ohio: Merrill, 1989], p. 109). See also Dinsmoor, "In the Beginning," esp. 293–94.
26. Clark L. Hull to BFS, Jan. 8, 1934, HA.
27. Skinner remembered that "Hull coined the term 'Skinner Box' and adopted my methods, but . . . repeatedly pass[ed] over my papers. Why? Was this characteristic of people in 'learning'?" (BFS, "The Young Man," May 2, 1972, BA).
28. BFS, "Hull's *Principles of Behavior*," *American Journal of Psychology* 57 [1944]: 276–81; BFS, "Courage," Jan. 8, 1965, BA; Dinsmoor, "In the Beginning," 293–94.
29. BFS to Clark L. Hull, Jan. 4, 1936, HA.
30. Clark L. Hull to BFS, Jan. 8, 1936, HA.
31. Leonard Carmichael to Carl Murchison, n.d., HA.
32. BFS to S. S. Stevens, June 16, 1935, HA.
33. BFS to S. S. Stevens, Sept. [?], 1934, HA.
34. BFS, "The Young Man," May 2, 1972, BA.
35. EGB to BFS, Dec. 2, 1946, HA.
36. BFS "Self Confidence," Nov. 19, 1960, BA. He also recalled that for perhaps fifteen years he worked in relative isolation. "It wasn't until the war was over that other psychologists paid any attention to my work . . . because the war came on within a couple of years of the publication of *Behavior of Organisms*" (Interview with BFS, Dec. 14, 1989).
37. BFS to his parents, Dec. 1929, BA.
38. BFS to his parents, Jan. 1930, BA.
39. Interview with BFS, Mar. 7, 1990.
40. Ibid.
41. Ibid.
42. Interview with BFS, July 13, 1988.
43. BFS, *Shaping of a Behaviorist,* p. 137.
44. Ibid.
45. Ibid., p. 145.
46. Ibid., p. 145.
47. BFS, "Honor," Sept. 28, 1966, BA.
48. Interview with YBS by Rhonda K. Bjork, July 6, 1989. Yvonne Blue said of her maternal grandfather, Opie Read, "[He is] my idol. No one has meant more to

me" (Yvonne Blue to Margaret [Artman], n.d., Yvonne Blue Skinner Collection, Arthur and Elizabeth Schlesinger Library on the History of Women in America, RC).

49. Yvonne Blue's Jan.–June 1925 diary, Mar. 1, 1925, RC.
50. Interview with BFS, Dec. 12, 1989.
51. Yvonne Blue's 1926–1937 diary, Mar. 24, 1937, RC.
52. Interview with BFS, July 10, 1989.
53. BFS, *Shaping of a Behaviorist,* p. 188.
54. Grace B. Skinner to BFS, n.d., BA.
55. Interview with BFS, July 10, 1989.
56. Ibid.
57. Yvonne Blue's 1926–37 diary, Mar. 24, 1937, RC.
58. Interview with BFS, July 10, 1989.
59. Interview with BFS, Dec. 12, 1989.
60. Interview with YBS, Jul. 6, 1989; *Cosmopolitan,* Aug. 1971, p. 80, HA.
61. Interview with BFS, Dec. 12, 1989.
62. Ibid., July 10, 1989.
63. BFS, untitled note, Jan. 1, 1970, BA.
64. EGB to BFS, Apr. 8, 1936, HA.
65. EGB to Richard M. Elliott, June 11, 1936, UM.
66. Richard M. Elliott to EGB, June 26, 1936, UM.
67. Leonard Carmichael to BFS, June 23, 1936, HA.
68. Walter S. Hunter to BFS, June 17, 1936, HA.
69. Richard M. Elliott to BFS, June 18, 1936, UM.
70. Paul E. Meehl to BFS, July 19, 1977, HA.
71. Kenneth MacCorquodale to BFS, July 12, 1977, HA.
72. "Progress in Psychology at Minnesota (1890–1953)," unpublished manuscript, n.d., UM. Author unknown.
73. BFS to William J. Crozier, May 20, 1937, HA.
74. Interview with BFS, Dec. 14, 1989.
75. Interview with BFS, Mar. 9, 1990.
76. BFS to Richard M. Elliott, Dec. 17, 1963, HA.
77. This plan was implemented from 1937 to 1945. See Richard M. Elliott, *A History of Psychology in Autobiography,* vol. 4 (Worcester, Mass.: Clark University Press, 1952), pp. 76–90, 89–90.
78. Speaking of his time as a Junior Fellow, Skinner said, "I had a large 'book' composed of sheets of card-board on which I wrote or posted formulae for verbal responses—speaker and listener. It was mainly a structural enterprise. Verbal stimuli—no verbal responses; non-verbal stimuli—verbal responses; verbal stimulus; verbal responses, and so on. I was still speaking of reflexes—pre-operant analysis. . . . I don't remember when I began to get the operant formula into it" (BFS, "V. B.," Dec. 20, 1973, BA).
79. BFS, "Has Gertrude Stein a Secret?" *Atlantic Monthly* 153 (1934): 50–57.

80. See BFS, "The Verbal Summator and a Method for the Study of Latent Speech," *Journal of Psychology* 2 (1936): 71–107. The story of the verbal summator is also told in *The Shaping of a Behaviorist,* pp. 174–76.
81. See BFS, "The Alliteration in Shakespeare's Sonnets: A Study of Literary Behavior," *Psychological Record* 3 (1939): 186–92.
82. BFS, *Verbal Behavior* (New York: Appleton-Century-Crofts, 1957).
83. Much of Skinner's rat work at Minnesota was conducted with a colleague, W. T. Heron, who was interested in using multiple animals and determining the effects of drugs. Several articles resulted, but Skinner had trouble coordinating the use of twenty-four lever boxes for one experiment (the old problem of variability), so he returned to cumulative curves on single animals. See the following articles that Skinner co-authored with Heron: "Effects of Caffeine and Benzedrine upon Conditioning and Extinction," *Psychological Record* 1 (1937): 340–46; and "An Apparatus for the Study of Animal Behavior," *Psychological Record* 3 (1939): 166–76.
84. BFS, "Pigeons in a Pelican," *American Psychologist* 15 (1960): 28–57. The historian of science James H. Capshew allowed me to read his forthcoming article, which argues persuasively that Skinner's shift of attention from the laboratory to "real life" began with Project Pigeon. See Capshew, "Engineering Behavior: Project Pigeon, World War II, and the Conditioning of B.F. Skinner," forthcoming in *Technology and Culture*, fall 1993.
85. BFS, *Shaping of a Behaviorist,* p. 241.
86. Ibid., p. 242.
87. BFS to Robert W. Powell, Feb. 4, 1971, HA.
88. Norman Guttman to BFS [ca. 1976], HA.
89. BFS, *Shaping of a Behaviorist,* pp. 267–68, 274.
90. John Tate to R. C. Tolman, Feb. [?], 1942, quoted in BFS, "History of the 'Pigeon Project' Contract with NDRC," unpublished manuscript, n.d., HA.
91. Richard M. Elliott to EGB, Jan. 9, 1946, UM.
92. BFS, " 'Pigeon Project' Contract."
93. Ibid.
94. Ibid.
95. Ibid.
96. Spencer Klaw, "Harvard's Skinner, The Last of the Utopians," *Harper's,* Apr. 1963, pp. 45–51, 48.
97. Norman Guttman to BFS, Feb. 18, 1976, HA.
98. Thomas P. Hughes, *American Genesis: A Century of Invention and Technological Enthusiasm* (New York: Viking Press, 1989), pp. 96–137.

CHAPTER 6

1. Interview with YBS by Rhonda K. Bjork, July 6, 1989.
2. Interview with BFS, Dec. 12, 1989. [Julie Skinner Vargas was present and occasionally participated.]

3. BFS, "The First Baby Tender," unpublished manuscript, n.d., BA.
4. BFS, *The Shaping of a Behaviorist* (New York: Knopf, 1979), pp. 274–75. See also "Brief Description [of the aircrib]," unpublished manuscript [1949 or 1950?], HA.
5. BFS, *Shaping of a Behaviorist,* pp. 275–76.
6. See James B. Gilbert, *Another Chance: Postwar America, 1945–1985* (New York: Dorsey Press, 1986), pp. 54–75. Landon Jones has an interesting account of the baby boom in *Great Expectations: America and the Baby Boom Generation* (New York: Ballantine, 1986).
7. Steven Mintz and Susan Kellogg, *Domestic Revolutions: A Social History of American Family Life* (New York: Free Press, 1988), pp. 133–201.
8. Carl N. Degler, *At Odds: Women and the Family in America from the Revolution to the Present* (New York: Oxford University Press, 1980), p. 9.
9. Mintz and Kellogg, *Domestic Revolutions,* pp. 188–90.
10. BFS to Cuthbert and Janet Daniel, Mar. 15, 1945, HA.
11. A. E. Bennett to Messrs. Hyde, Kuphal, and Graham, Oct. 27, 1944, HA.
12. BFS to Mary Lea Page, June [?], 1945, HA.
13. Mary Lea Page to BFS, July 9, 1945, HA.
14. BFS, "Baby in a Box," *Ladies' Home Journal,* October 1945, pp. 30–31; 135–136; 138.
15. Mary Seth to BFS, Nov. 21, 1945, HA. [Ms. Seth quoted Skinner on the number of tenders being built.]
16. BFS, "Baby in a Box," in *Readings in Developmental Psychology,* ed. Judith Krieger Gardner (New York: Little, Brown, 1978), p. 98. (This quote is from Skinner's introduction to the reprint of "Baby in a Box," 98–104.
17. "A Reader of the Times" to the district attorney of Bloomington, Ind., Sept. 30, 1945, HA.
18. Interview with BFS, Dec. 12, 1989.
19. " 'Aircrib' Experience as Toddler was 'marvelous,' woman declares." Interview with Deborah Skinner Buzan, by Oz Hopkins, *Oregon Journal,* Mar. 10, 1975, HA.
20. Interview with Deborah S. Buzan, Feb. 8, 1993.
21. BFS to Charles S. Waugh, July 22, 1974, HA.
22. BFS, "First Baby Tender."
23. John H. Gray to BFS, Aug. 1, 1946, HA.
24. BFS to Mary Seth, Nov. 30, 1945, HA.
25. BFS to Mrs. Armand Denis, Oct. 18, 1945, HA.
26. Charles H. Stearns to the *Ladies' Home Journal,* July 6, 1946, HA.
27. Harriet Read Blue to BFS, Sept. 27, 1945, HA.
28. BFS to Mary Seth, Nov. 30, 1945, HA.
29. JWJ to BFS, Oct. 15, 1945, HA.
30. H. B. Wells to BFS, Dec. 19, 1944, HA. For a brief discussion of Skinner at Indiana University, see *Psychology at Indiana University: A Centennial Review and Compendium*, ed. Eliot Hearst and James H. Capshew (Bloomington: University of Indiana Press, 1988), pp. 55–60.

31. Interview with YBS, July 6, 1989.
32. BFS to JWJ, Oct. 15, 1945, HA.
33. Ibid.
34. JWJ to BFS, Oct. 19, 1945, HA.
35. BFS to JWJ, Oct. 15, 1945, HA.
36. Ibid.
37. JWJ to BFS, Oct. 19, 1945, HA.
38. George Green to YBS, Oct. 20, 1945, HA.
39. BFS to JWJ, Nov. 17, 1945, HA.
40. Ibid.
41. BFS to JWJ, Nov. 24, 1945, HA.
42. H. J. Glickman to BFS, Dec. 6, 1945, HA.
43. BFS to JWJ, Mar. 23, 1946, HA.
44. Ibid.
45. L. J. Schwartz to BFS, July 16, 1948, HA.
46. BFS to Daniel E. Caldemeyer, May 24, 1946, BA.
47. BFS to Daniel E. Caldemeyer, June 14, 1946, BA.
48. BFS to Daniel E. Caldemeyer, Oct. 17, 1946, BA.
49. Daniel E. Caldemeyer to BFS, Aug. 18, 1947, BA.
50. Interview with BFS, Mar. 5, 1990.
51. For an interesting discussion of new directions in a consumer-oriented postwar economy, see Erik Barnouw, *Tube of Plenty: The Evolution of American Television* (New York: Oxford University Press, 1982); and John Kenneth Galbraith, *The Affluent Society* (New York: Houghton Mifflin, 1958).
52. BFS to Laurence Kingsland, Sept. 14, 1948, HA.
53. BFS to Richard Moss, Mar. 20, 1951, HA.
54. BFS to John M. Gray, Oct. 28, 1952, HA.
55. A promotional flyer for John Gray's aircrib, HA.
56. BFS to Mimi Meyer, July 25, 1968, HA.
57. Among the articles continuing to describe the aircrib were the following: "Baby Box," *New Yorker,* July 19, 1947, pp. 19–20; "Box reared Babies: Skinner Baby Box," *Time* 64, Feb. 2, 1954, pp. 66; Berkeley Rice, "Skinner Agrees He Is Most Important Influence in Psychology," *New York Times Magazine* Mar. 17, 1968, p. 27.
58. BFS, *A Matter of Consequences,* p. 251.
59. Interview with YBS, July 6, 1989.
60. For more on America's love affair with suburbia, see Kenneth Jackson, *Crabgrass Frontier: The Suburbanization of America* (New York: Oxford University Press, 1985); Scott Donaldson, *The Suburban Myth* (New York: Columbia University Press, 1969); and Herbert Gans, *The Levittowners* (New York: Pantheon, 1967).
61. BFS to Donald Bullock, Feb. 5, 1960, HA.
62. BFS to Hilda Hunter, Jan. 25, 1957, HA.
63. James A. Dinsmoor, "A Visit to Bloomington: The First Conference on the

Experimental Analysis of Behavior," *Journal of the Experimental Analysis of Behavior* 48 (1987): 441–45. Quote is from James A. Dinsmoor to Daniel W. Bjork, June 12, 1992, author's possession.

64. BFS to John P. Gluck, Mar. 29, 1973, HA.
65. *Indiana Daily Student,* n.d., HA.
66. Dinsmoor, "A Visit to Bloomington," p. 444.
67. BFS to Arthur Bentley [Summer 1947], HA.
68. James A. Dinsmoor, "Keller and Schoenfeld's *Principles of Psychology," Behavior Analyst* 12 (1989): 213–19.
69. Transcript of Skinner film sound track. n.d., HA. (FSK was also present.)
70. William N. Schoenfeld, "Reminiscences You Say," *Journal of the Experimental Analysis of Behavior* 48 (1987): 464–68.
71. BFS to Wilse Webb, Feb. 11, 1960, HA.
72. Victor G. Laties, "Society for the Experimental Analysis of Behavior: The First Thirty Years (1957–1987)," *Journal of the Experimental Analysis of Behavior* 48 (1987): 495–512.
73. Interview with YBS, July 6, 1989, by Rhonda K. Bjork.
74. Interview with YBS, Dec. 19, 1988, by William R. Woodard, University of New Hampshire.
75. BFS, "My Years at Indiana" (Paper prepared for the Indiana Psychology Centennial Celebration, Apr. 8, 1988). Skinner was unable to attend the celebration, but he sent the paper. James Dinsmoor was kind enough to share a copy with the author. Celebration took place in Bloomington.

CHAPTER 7

1. BFS, *The Shaping of a Behaviorist* (New York: Knopf, 1979), p. 292.
2. BFS to Toni Kross, July 13, 1964, HA.
3. Alice Felt Tyler, *Freedom's Ferment: Phases of American Social History from the Revolution to the Outbreak of the Civil War* (1944; reprint, New York: HarperCollins, 1962). Tyler, a history professor at the University of Minnesota, had given Skinner a copy of her recently published book. One correspondent asked specifically if Skinner had studied utopian literature before writing *Walden Two* to gain a sense of shaping a community through education. He replied: "I did poke about in the history of American communities, but never built up a formal bibliography." He did, however, mention several other sources. "I believe there is something in Bacon's *New Atlantis* and of course Plato and More." He said, "[T]he accent on nineteenth century Utopian books [was] mainly economic. I do not recall anything about education in Morris's *News from Nowhere* or Butler's *Erewhon,* or in Bellamy's *Looking Backward"* (BFS to Jack Vernon, Nov. 4, 1955, HA).

   An astute student of Skinner's intellectual antecedents, especially his indebtedness to Francis Bacon, has recently noted that a belief in the power of a new technology to build a radically better society was central to both *Walden Two* and Bacon's *New Atlantis* (Laurence D. Smith, "On Prediction and Control: B. F. Skinner and the Technological Ideal of Science," *American Psychologist* 47 [Feb. 1992]: 216–23). For a book-length discussion of technology in American utopian ventures, see Howard P. Segal, *Technological Utopianism in American Culture* (Chicago: University of Chicago Press, 1985).

4. BFS, *Shaping of a Behaviorist,* p. 293.
5. BFS, "Walden Two," Feb. 17, 1987, BA.
6. BFS to Toni Kross, July 13, 1964, HA.
7. BFS, "Input-Output," Mar. 24, 1970, BA.
8. BFS, *Shaping of a Behaviorist,* p. 296.
9. BFS, *Walden Two* (New York: Macmillan, 1948), p. 30.
10. BFS, "Walden Two Revisited," in *Walden Two* (reissued ed., New York: Macmillan, 1976), p. xiii.
11. BFS, *Walden Two,* p. 193.
12. Ibid., pp. 255–56.
13. Ibid., pp. 256–57.
14. BFS, *Shaping of a Behaviorist,* p. 296. These discussions also helped Skinner fashion the novel's extensive and often combative dialogue. Of the group, Skinner was most intimate with Herbert Feigl, who was a member of the 1930s positivist group "the Vienna Circle." "Feigl . . . and I became close friends," he recalled, "although as he [Feigl] put it, we continued to 'cultivate our own gardens.' " He also said, "[Feigl] was one of a group of friends who came once a week to hear parts of *Walden Two* as I wrote them." He did not mention whether the others in this "group of friends" were also the ones who had originally helped stimulate his thinking (BFS, *Recent Issues in the Analysis of Behavior* [Columbus, Ohio: Merrill, 1989], p. 107).
15. BFS, "Walden Two Revisited," p. v.
16. BFS to Wade Van Dore, Apr. 21, 1969, HA.
17. Interview with BFS, Dec. 19, 1988.
18. Ibid.
19. "B. F. Skinner," *Miami Herald,* Jan. 27, 1979, HA.
20. BFS to Yolande Tremblay, July 28, 1970, HA.
21. BFS to Donald C. Williams, Oct. 4, 1949, HA.
22. BFS, "Nom de Plume," Apr. 3, 1961, BA.
23. BFS, *Shaping of a Behaviorist,* pp. 297–98.
24. BFS to John McCreary, Oct. 8, 1948, HA.
25. Ibid.
26. Richard M. Elliott to EGB, Jan. 9, 1946, UM.
27. BFS to Stephan Feinstein, Oct. 7, 1969, HA.
28. BFS, "Unconscious," Jan. 22, 1976, BA.
29. BFS, "Non Sub Homine," Nov. 29, 1973, BA.
30. BFS to Roberta Finn, Sept. 25, 1970, HA.
31. The dedication read "To W. A. S. and G. B. S."
32. BFS, "Who Will Control? Who Controls the Controller?" Feb. 9, 1968, BA.

33. Richard M. Elliott to EGB, Jan. 9, 1946, UM.
34. BFS to Elizabeth Green, June 8, 1950, HA. The quotation was reportedly from Russell Maloney of CBS's "Of Men and Books."
35. *New York Herald Tribune,* June 9, 1948, HA.
36. *Chicago Sunday Tribune,* June 13, 1948, HA.
37. Charles Poore, "Tour of an Almost Perfect Union," *New York Times Book Review,* p. 6; June 13, 1948, and June 11, 1948, HA.
38. *New Yorker,* June 12, 1948, HA.
39. John Hutchens to BFS, Apr. 26, 1948, HA.
40. Charles Anderson to BFS, June 11, 1948, HA.
41. "The Newest Utopia," *Life,* June 28, 1948, p. 38, HA.
42. "Utopia Bulletin," *Fortune,* Oct. 1948, pp. 191–96.
43. "The Newest Utopia," p. 38.
44. "Utopia Bulletin," p. 196.
45. BFS, *Shaping of a Behaviorist,* p. 348.
46. BFS to editor, *Life,* July 1, 1948, HA.
47. BFS to Paula Dietrichson, Nov. 3, 1966, HA.
48. J. P. Cuyler to BFS, Aug. 19, 1948, HA.
49. BFS to Claire S. Degner, Oct. 3, 1968, HA.
50. Joseph W. Krutch, *The Measure of Man: On Freedom, Human Values, Survival and the Modern Temper* (New York: Grosset & Dunlap, 1954), pp. 32–33.
51. Ibid., p. 57.
52. Ibid., p. 64.
53. Ibid., p. 65.
54. Ibid., pp. 68–71.
55. BFS to Richard M. Elliott, Mar. 13, 1955, UM.
56. BFS to Karl Meyer, July 6, 1954, HA.
57. BFS, "Freedom and the Control of Men," *American Scholar* 25 (Winter 1955–1956): 47–65.
58. See BFS and Carl Rogers, "Some Issues Concerning the Control of Human Behavior: A Symposium," *Science* 124 (1956): 1057–66.
59. BFS, untitled note [1948], HA. The bulk of his handwritten notes are at the HA but have not been prepared for scholarly use. Typed copies of most of the original notes are in the BA at the Skinner Cambridge residence.
60. BFS to Charles P. Curtis, Mar. 20, 1949, HA.
61. BFS, *A Matter of Consequences,* p. 254.
62. BFS to Matthew Israel, Aug. 16, 1956, HA.
63. BFS to Arthur Gladstone, Oct. [?], 1955, HA.
64. BFS, untitled note [1955?], BA.
65. BFS, *A Matter of Consequences,* p. 78.
66. BFS, untitled note [spring 1955], quoted in BFS, *A Matter of Consequences,* pp. 78–79.
67. BFS, "Statements of Principles," "Responsibility," "Problem of Productive Use of Time," "What the Community Guarantees the Individual," "Financial

Sanctions," "A Name," "Receiving a New Child in the Community," "The Problem of Cleanliness," and "The Contented Cow." No specific dates appear on any of these notes, but all were written between January and June of 1955 at Putney, Vt., and all are in the BA.

68. BFS, *A Matter of Consequences,* p. 84.
69. Richard M. Elliott to EGB, Jan. 9, 1946, UM.
70. The first and only issue was *Walden Two Bulletin,* August 20, 1956, HA.
71. BFS to Alan Levensohn, Apr. 14, 1959, HA.
72. BFS, *A Matter of Consequences,* p. 252.
73. For an excellent sampling of counterculture writing, see *The Sixties: Art, Politics and Media in Our Most Explosive Decade,* ed. Gerald Howard (New York: Paragon House, 1991).
74. Stephanie Jean Hemphill to BFS [early 1960s?], HA.
75. Mary Ann Sims to BFS, Mar. 28, 1968, HA.
76. Betty Ward to BFS, Apr. 29, 1972, HA.
77. Pamela Myers to BFS, May 22, 1967, HA (her emphasis).
78. Transcript of Yorkshire Film [1977], HA. This film became a BBC television special on Skinner, aired in the spring of 1977. Skinner was not particularly happy with the production. Later WGBH (Boston) financed a "Nova" special on BFS.
79. Helen Derevnuk to BFS, Aug. 18, 1969, HA.
80. R. K. Niskanen to BFS, July 17, 1963, HA.
81. There were two short-lived Skinnerian-type communities established in 1966 in urban areas—"Walden Pool," in Atlanta, and "Walden House," in Washington, D.C. See BFS to William Shepard, Nov. 28, 1966, HA; and Joseph William Majewsky to BFS, Mar. 21, 1966, HA. Also, a group of radical behaviorists at Western Michigan University, led by the former chairperson of the department of psychology, Roger Ulrich, helped establish Experimental Community Two, or "EC2," and a "Learning Village," both of which used reinforcement in education. Something of EC2 remains on the shores of a small lake outside of Kalamazoo, but it is no longer really a Walden Two–type community. For a flavor of EC2, see copies of the 1969 *EC2 Newsletter* in HA.

    Skinner corresponded frequently with Walden Two enthusiasts and encouraged the efforts just mentioned, as well as several others that did not survive—like "East Wind," a community in Missouri. He never, however, joyfully joined the counterculture, as did Timothy Leary, who once asked Skinner why he did not introduce drugs into Walden Two communities. For Skinner's reaction to the emerging counterculture interest in *Walden Two,* see related correspondence in the Harvard Archives; and BFS, *A Matter of Consequences,* pp. 306–8. "Quite honestly," Skinner admitted, "I was never one of the gurus of the sixties" (ibid., p. 307).
82. A short description of Twin Oaks appears in *Time,* "Skinner's Utopia: Panacea or Path to Hell?" September 20, 1971, pp. 48–49; and a more extensive and

recent description of Los Horcones—which follows Skinnerian techniques closely—appears in Steve Fishman, "The Town B. F. Skinner Boxed," *In Health* Jan./Feb. 1991, pp. 50–60. It is also interesting to read *The Leaves of Twin Oaks,* the Twin Oaks newsletter, copies of which are in the Harvard Archives.

Skinner may not have been a guru, but others made him one. "Josh" (community members usually dropped their last names) paraphrased a biblical parable: "And it came to pass that a great nation rose up and held dominion over other nations and oppressed them sorely and this oppression was called freedom. . . . And these people did worship the god Money and did build mighty temples unto him which they called banks. . . . But in the place called Harvard there arose among the scribes and wise men a prophet named B. F. Skinner and Skinner did speak unto the elders and scribes, saying, 'Your teachings are false. Listen unto me and I will tell thee of the science of human behavior.' And they did scoff at him and mock him but he did keep his cool and spake again unto them saying, 'Thou lackest understanding.' " (*The Leaves of Twin Oaks,* Oct. 1970, pp. 11–12.)

83. See the newsletter of Los Horcones, *Walhdos 40;* and "The Experimental Analysis of Behavior Applied to Communal Living: Los Horcones Community," n.d., HA.
84. BFS to Patrick C. Burns, Nov. 21, 1967, HA.
85. BFS to Lisa J. Stragliotto, Nov. 16, 1979, HA.
86. BFS to Kathleen Kincade, May 3, 1977, HA. Roger Ulrich, who started the "Learning Village" in Kalamazoo, was instrumental in persuading Skinner to make the visit.
87. BFS to Patrick C. Burns, Nov. 21, 1967, HA.
88. Skinner rarely praised other books. An exception was the "remarkable" *Small Is Beautiful,* by the British economist E. F. Schumacher ("Walden Two Revisited," pp. ix–x). Mass society and the prospect of uncontrollable population growth made it less and less likely that behavioral technology, which had always worked best with the individual or small group, would be effective.
89. BFS interview for the *Kentucky Kernel* (University of Kentucky), Nov. 12, 1965, HA.
90. Richard Graham (Dandelion Community, Enterprise, Ontario, Canada) to BFS, Feb. 5, 1975, HA.
91. BFS to Robert Michaels, Sept. 11, 1967, HA.
92. Skinner's last published writing on *Walden Two* was an imaginative piece called, like William Morris's utopian book, "News from Nowhere," but Skinner added a chronological twist by tacking on "1984" at the end of the title. He imagined that George Orwell faked his own death; came to live in Walden Two under his birth name, Eric Arthur Blair; and got into an extended discussion with Frazier, who passionately defended "programmed instruction." See "News from Nowhere, 1984," in *Upon Further Reflection* (Englewood Cliffs, N.J.: Prentice-Hall, 1987), pp. 42–43. Discussion about specially designed communities

like Walden Two and "programmed instruction" was, in effect, discussion about the application of behavioral engineering to two different environments—the village and the metropolis.

CHAPTER 8

1. EGB to Richard M. Elliott, Jan. 5, 1946, UM.
2. The ten William James Lectures on Psychology were as follows: Oct. 10, "The Age of Words"; Oct. 17, "Verbal Behavior as a Scientific Subject Matter"; Oct. 24, "Types of Verbal Behavior"; Oct. 31, "Words and Things: The Problem of Reference"; Nov. 7, "Multiple Sources of Verbal Strength"; Nov. 14, "Making Sentences"; Nov. 21, "The Effect Upon the Listener"; Nov. 28, "Understanding: Real and Spurious"; Dec. 5, "Thinking in Words"; Dec. 12, "The Place of Verbal Behavior in Human Affairs."
3. EGB to Richard M. Elliott, Dec. 13, 1947, UM.
4. BFS, *Shaping of a Behaviorist,* p. 340.
5. Interview with BFS, Dec. 12, 1989.
6. Interview with YBS by William R. Woodward, Dec. 19, 1988.
7. BFS, *Shaping of a Behaviorist,* p. 341.
8. BFS, "Summer of 1941" [late 1950s or early 1960s?], BA. Of course, Julie could have been expressing mere jealousy.
9. See YBS, "Record of Julie's Speech During a Twenty Minute Period," unpublished note, May 1, 1941, HA. At this time Skinner was recording his own verbal production. He used a timer to record the hours he spent at his desk composing, and he even counted the number of words he wrote in a given period.
10. BFS, "Debbie's Verbal Behavior," Jan. 11, 1959, BA.
11. BFS, "Proofreading" [1960?], BA.
12. Interview with BFS, Mar. 9, 1990. Julie Skinner Vargas participated.
13. BFS, "My Instructions to Eve," unpublished manuscript, (1944–1945), HA.
14. BFS, "Memorandum to Members of the Department of Psychology [Fall 1955], HA.
15. BFS, *A Matter of Consequences* (New York: Knopf, 1984), p. 64.
16. BFS to Glenna Hyde, Nov. 1, 1966, HA.
17. Interview with Richard Herrnstein, Aug. 14, 1990.
18. BFS, *Matter of Consequences,* p. 65.
19. Interview with Deborah Skinner Buzan, Feb. 8, 1993.
20. For a good description of these early teaching machines, see Edward B. Fry, *Teaching Machines and Programmed Instruction: An Introduction* (New York: McGraw-Hill, 1963), pp. 19–20. See also Skinner's article, "Teaching Machines," *Science,* October 24, 1958, pp. 969–77. The most comprehensive book on both teaching machines and programming in the early years is A. A. Lumsdaine and Robert Glaser, *Teaching Machines and Programmed Learning: A Source Book* (Washington, D.C.: National Education Association, 1960).

21. Donald A. Cook recently made this important point in a letter to Daniel W. Bjork, Dec. 6, 1992; author's possession (no collection). Cook, who had a doctorate from Columbia University, worked on programmed instruction in the early years and impressed Skinner with his skill.
22. Ernest A. Vargas and Julie S. Vargas, "Programmed Instruction and Teaching Machines," in *Designs for Excellence in Education,* ed. Richard P. West and L. A. Hamerlynack (Longmont, Colo.: Sopris West, 1992), p. 36.
23. For a comparison of Pressey's and Skinner's teaching machines, see Edward Fry, "Teaching Machine Dichotomy: Skinner vs. Pressey," in *Programmed Learning: Theory and Research,* ed. Wendell I. Smith and J. William Moore (Princeton, N.J.: Van Nostrand, 1962), pp. 81–86. Pressey's description of his first teaching machine was published as "A Simple Device for Teaching, Testing, and Research," *School and Society* 23 (1926): 373–76.
24. BFS, "The Science of Learning and the Art of Teaching," *Harvard Educational Review* 24 (1954): 90.
25. Ibid., pp. 96–97.
26. BFS to Gabriel D. Ofiesh, Mar. 12, 1970, HA.
27. BFS, *Matter of Consequences,* pp. 95–96.
28. Ibid., p. 96. For a technical, yet broadly conceived, discussion of prompting in various verbal environments, see chap. 10, "Supplementary Stimulation," in BFS, *Verbal Behavior* (New York: Appleton-Century-Crofts, 1957), pp. 253–92.
29. BFS, *Matter of Consequences,* pp. 96–97.
30. James Holland and BFS, *The Analysis of Behavior* (New York: McGraw-Hill, 1961).
31. BFS, *Matter of Consequences,* p. 119.
32. *Harvard Alumni Bulletin* [May 1960], p. 638, HA.
33. BFS, *Matter of Consequences,* p. 185.
34. Ibid., p. 71. He did get financial support from the Ford Foundation.
35. BFS, "Company's Misbehavior" [1963], BA.
36. Fry, *Teaching Machines,* p. 20.
37. Transcript of interview with BFS for the National Society for Programmed Instruction [1972–1973], HA.
38. James L. Rogers to BFS [summer 1956]. Copy of letter sent to author.
39. BFS, *Matter of Consequences,* pp. 141, 158.
40. BFS, "Mistreatment," Nov. 19, 1986, BA.
41. BFS, *Matter of Consequences,* p. 159.
42. BFS to S. S. Stevens, July 28, 1959, HA.
43. S. S. Stevens to BFS, Aug. 7, 1959, HA.
44. Edwin B. Newman to BFS, July 29, 1959, HA.
45. BFS, *Matter of Consequences,* p. 159.
46. BFS to C. V. Coons, May 11, 1961, HA (his emphasis).
47. C. V. Coons to BFS, May 16, 1961, HA.
48. BFS to C. V. Coons, June 26, 1961, HA.
49. C. V. Coons to BFS, July 31, 1961, HA.

50. Ibid.
51. BFS, "Stock Taking," Apr. 21, 1961, BA.
52. BFS, "Early Thought," Feb. 28, 1962, BA.
53. BFS, "Business," Nov. 20, 1962, BA.
54. See Robert Wiebe, *The Search for Order: 1877–1920* (New York: Hill & Wang, 1967); and Gabriel Kolko, *The Triumph of Conservatism* (New York: Free Press, 1963). Wiebe and, especially, Kolko illustrate how American corporations attempted to mitigate risky business conditions for steadier, if less spectacular, profits. Survivability and steady growth, not quick money, were the ideals in large American business after the Civil War and throughout the twentieth century.
55. BFS, "Educational Specialists," June 17, 1966, BA.
56. James B. Conant, *The American High School Today: A First Report to Interested Citizens* (New York: McGraw-Hill, 1959).
57. For an entertaining and insightful discussion of the American reaction to *Sputnik,* see William Manchester, *The Glory and the Dream: A Narrative History of America 1932–1972* (Boston: Little, Brown, 1974), pp. 787–814.
58. BFS, *Matter of Consequences,* pp. 38, 166.
59. Ibid., p. 91.
60. Pauly, *Controlling Life: Jacques Loeb and the Engineering Ideal in Biology* (New York: Oxford University Press, 1987), pp. 90–91.
61. Skinner erroneously dated the meeting December 31, 1958—a year early—in *A Matter of Consequences,* p. 166.
62. BFS to James B. Conant, Jan. 14, 1960, HA.
63. BFS to I. G. "Jack" Davis, Jr., Jan. 4, 1960, HA.
64. James B. Conant to BFS, Jan. 18, 1960, HA.
65. Ibid.
66. BFS to James B. Conant, Feb. 4, 1960, HA.
67. BFS to I. G. "Jack" Davis, Jr., Feb. 11, 1960, HA.
68. Several chapters of the *Technology of Teaching* had been published previously, between 1954 and 1965. See BFS, the *Technology of Teaching* (New York: Appleton-Century-Crofts, 1968): vii.
69. BFS, "Organum," Sept. 19, 1966, BA (his emphasis).
70. Quoted in John D. Green, "B. F. Skinner's Technology of Teaching," *Classroom Computer Learning* (Feb. 1984): 24.
71. Donald A. Cook shared this remark he recalled Skinner making with the author. Donald A. Cook to Daniel W. Bjork, Dec. 6, 1992.
72. Green, "Skinner's Technology of Teaching," p. 28.
73. For a current analysis of the computer as a teaching machine, see Donald A. Cook, "Can Computers Help?" *CBT Directions* 5 (July/Aug. 1992): 9–13; and Vargas and Vargas, "Programmed Instruction," pp. 50–52.
74. Interview with BFS, Mar. 9, 1990. See also BFS, "Are Theories of Learning Necessary?" *Psychological Review* 57 (July 1950): 193–216.
75. Interview with BFS, Mar. 9, 1990.

76. BFS, "Writing", Oct. 21, 1968, BA.
77. BFS, "SOS," Feb. 10, 1960, BA.
78. BFS, "Depression," Sept. 19, 1961, BA.
79. Ibid.
80. Interview with YBS by William R. Woodward, Dec. 19, 1988.
81. BFS, "The Psycho City Saga: A Psychological Elizabethan Western" [1960], HA.
82. For example, Skinner recorded the following schedule of meetings dealing with the teaching machine and programmed instruction between February and July 1960: February 4, "Joint Meeting of Staff Members, Dana Hall School (Wellesley and Beaver Country Day School (Boston)"; February 9, "Organization of Graduate School of Education Students"; February 10, "Staff of the Boston City Hospital"; February 17, "American Association of School Administrators (Atlantic City)"; March 15, "American Textbook Publishers Institute (Philadelphia)"; April 7, "Child Psychiatrists and Pediatricians, Massachusetts General Hospital"; April 14, "Sandwich Seminar, Graduate School of Education (Cambridge)"; April 29, "School Psychologists Association (Nassau County, New York)"; May 6, "New England Board of Higher Education (Boston)"; May 10, "Deans and Other Officers of New York University (New York)"; May 11, "Group of Congressmen and Congresswomen at the Brookings Institution (Washington)"; May 21, "Massachusetts Association of School Boards (Lenox, Massachusetts)"; June 29, "Group of College Presidents, Harvard Business School"; July 8, "School Administrators Conference, Spaulding House, MIT (BFS, datebook, BA).

    Skinner also noted, "During this period [Feb.–July] more than twenty-five groups of people visited the [Harvard] Laboratories for discussions of the teaching machines, including several from foreign countries. Interviews were given several people preparing articles for national publications, a paper had been prepared for *The Scientific American* and the teaching machines were made the focal point of a television program on the *Conquest* Program of CBS" ("Lectures and Discussions of Teaching Machines," unpublished manuscript, n.d. [almost certainly 1960, because of mention of the "Conquest" program]), HA. The next year he went to the Soviet Union to lecture on—of course—teaching machines.
83. BFS, "Mitigations of These Pessimistic Comments," July 13, 1963, BA.
84. Interview with BFS, Dec. 14, 1989.
85. For instance, Skinner recorded:

> [A] quiet verbal 'rumble' between Bruner and me [occurred] at a recent meeting of the Committee on Programmed Instruction. We were discussing possible activities on a national scale. I brought up the possibility of working jointly with some of the school curriculum committees. Bruner said in essence, 'Yes, since programmers have been so short on subject-matter specialization.' I rushed right in: 'And the curriculum committees have been short on how to teach.' I'm afraid

I started it. The theme of the above note was in the air. The note was written the morning of the day the committee met.

(BFS, "The School Curriculum Committees," Dec. 12, 1963, BA.)

86. BFS, "Stock Taking," [1961], BA.
87. BFS, "My Day," Aug. 9, 1963, BA.

CHAPTER 9

1. *Time,* Sept. 20, 1971, pp. 47–53.
2. BFS, "The Motivation of B. F. S.," Sept. 12, 1971, BA. (The *New York Times* article Skinner referred to was Berkeley Rice, "Skinner Agrees He Is the Most Important Influence in Psychology," *New York Times Magazine,* Mar. 17, 1968, p. 27.)
3. BFS to Max Tishler, Dec. 6, 1971; and BFS to Bert Decker, Dec. 11, 1972, HA.
4. Elizabeth Hall, "Will Success Spoil B. F. Skinner?" *Psychology Today,* Nov. 1972, p. 68.
5. Cover of paperback edition of *Beyond Freedom and Dignity* (New York: Bantam/Vintage, 1972).
6. BFS, "Science and Human Behavior," Jan. 1, 1970, BA.
7. BFS, *Beyond Freedom and Dignity* (New York: Knopf, 1971), p. 183.
8. Ibid., p. 206.
9. See "Freedom and the Control of Men," *American Scholar* 25 (Winter 1955–1956): 47–65; BFS and Carl Rogers, "Some Issues Concerning the Control of Human Behavior: A Symposium," *Science* 124 (Nov. 30, 1956): 1057–66.

   For a good example of Skinner's thinking during his interaction with scholars involved with the Fund for the Republic, see his paper entitled "The Concept of Freedom from the Point of View of a Science of Human Behavior," unpublished working paper for the Fund of the Republic, July 16, 1958, HA. Skinner noted that "freedom is not something the human species needs, as it needs food and water. The species has survived and even flourished even though only a small fraction of its members have known any substantial degree of political freedom" (ibid.).
10. BFS, "Culture," Feb. 3, 1968, BA (his emphasis).
11. BFS, "Geometry," May 14, 1972, BA.
12. BFS, "The Issues of the Day," May 9, 1971, BA (his emphasis).
13. BFS, untitled note, Feb. 22, 1972, BA.
14. James Dinsmoor to Daniel W. Bjork, June 12, 1992. Author's possession.
15. BFS, "Being Understood," Aug. 30, 1971, HA.
16. BFS, "The Aversive Consequences of Positive Reinforcement," Dec. 19, 1970, BA.
17. BFS, "Verbal Repertoire," Mar. 21, 1971, BA.
18. BFS, "Outlining," Oct. 2, 1964, BA
19. Ibid. (his emphasis).

20. This was published three years later as "The Design of Experimental Communities," *International Encyclopedia of the Social Sciences,* vol. 16 (New York: Macmillan, 1968), pp. 271–75.
21. BFS, "The Design of Cultures," Jan. 19, 1965, BA (his emphasis).
22. BFS, "Talcott Parsons," Oct. 28, 1969, BA. Parsons's comments were from *Societies: Evolutionary and Comparative Perspectives* (Englewood Cliffs, N.J.: Prentice-Hall, 1966), pp. 5, 7. The reviewer was Tom Bottomore, *New York Review of Books,* Nov. 6, 1969.
23. BFS, "Getting the Point," Nov. 24, 1968, and Feb. 23, 1971, BA.
24. BFS, "On Leaving Cambridge, England," May 28, 1969, BA.
25. BFS, *"DEPTH,"* May 12, 1969, BA (his emphasis).
26. BFS, "References," May 3, 1970, BA.
27. Ibid.
28. BFS, "The Failure of Experimental Psychology," May 28, 1970, BA.
29. BFS, "Lucky Hit," Apr. 7, 1971, BA.
30. Sidney R. Wilson to *Psychology Today,* Oct. 19, 1971, HA.
31. *TV Guide,* Oct. 17, 1971, HA.
32. Transcript of "Firing Line," recorded Oct. 2, 1971, HA.
33. Stewart Moore to BFS, Oct. 22, 1971, HA.
34. BFS to Stephen Kuffler, Oct. 28, 1971, HA.
35. BFS to Paul Meehl, Oct. 27, 1971, HA.
36. The *Times* quote was used on the cover of *The Chomsky Reader,* ed. James Peck (New York: Pantheon Books, 1987).
37. BFS, "My Critics", Dec. 23, 1971, BA. He amended, however, on Dec. 25, 1971, that he slept "well."
38. For analysis of Chomsky's linguistic system as well as short discussions on intellectual differences between Chomsky and Skinner, see Justin Leiber, *Noam Chomsky: A Philosophic Overview* (Boston: Twayne, 1975), pp. 140–44; and John C. Marshall, "Language Learning, Language Acquisition, or Language Growth?" in *B. F. Skinner: Consensus and Controversy,* ed. Sohan Modgil and Celia Modgil (New York: Falmer Press, 1987), pp. 41–43. There is no adequate biography of Chomsky, but see Chomsky's own discussion of his life in Peck, *The Chomsky Reader,* pp. 3–55.
39. BFS, *Beyond Freedom and Dignity,* pp. 28–30.
40. Ibid., pp. 205–6.
41. Peck, *Chomsky Reader,* p. 162.
42. Ibid., p. 163.
43. Ibid., p. 181.
44. BFS, "Chomsky," Jan. 30, 1972, BA.
45. BFS, "The Linguists," Jan. 10, 1968, BA (his emphasis). Though Skinner did not respond directly to Chomsky's review, others did. See, for example, Kenneth MacCorquodale, "On Chomsky's Review of Skinner's *Verbal Behavior," Journal of Experimental Analysis of Behavior* 13 (1970): 83–99.
46. Hall, "Will Success Spoil B. F. Skinner?" p. 68.

47. Ayn Rand, *Philosophy: Who Needs It?* (New York: Signet, 1984), pp. 137–38. Rand's review was written sometime between 1971 and 1973. See Leonard Peikoff's introduction in ibid., p. x.
48. Ibid., pp. 86–91.
49. Andrew A. Wallace to BFS, Sept. 24, 1971, HA.
50. The reviewer was Harriet Van Horne and the newspaper was probably a York, Pa., daily, since the letter that enclosed it had a York address. This, however, is inference, as the review newspaper was neither identified nor dated. See Van Horne, "Professor's Book Is Scary," HA.
51. Hall, "Will Success Spoil B. F. Skinner?" p. 68.
52. Two critical book-length studies inspired by negative reactions to *Beyond Freedom and Dignity* were Francis A. Schaeffer, *Back to Freedom and Dignity* (Cowners Grove, Ill.: Inter-Varsity Press, 1972); and H. Puligandla, *Fact and Fiction in B. F. Skinner's Science and Utopia* (St. Louis, Mo.: Green, 1974). One review that revived fears about Skinner's social inventions was Harold Kaplan's "Life in a Cage," *Commentary* 53 (Feb. 1972): 82–84, 86.
53. Ralph Waldo Emerson first used the cultural-political division "Party of Hope" and "Party of Memory." Emerson's insight is quoted and elaborated on in R. W. B. Lewis, *The American Adam: Innocence, Tragedy, and Tradition in the Nineteenth Century* (Chicago: University of Chicago Press, 1964), p. 7.
54. James M. McPherson, *Battle Cry of Freedom: The Civil War Era* (New York: Oxford University Press, 1988), p. vii.
55. Ralph Ketcham has a fine discussion of both the development of and contemporary dominance of American individualism in *Individualism and Public Life: A Modern Dilemma* (New York: Blackwell, 1987), pp. 1–70, 134–219.
56. Richard Sennett, "Review of *Beyond Freedom and Dignity,*" *New York Times Book Review,* Oct. 17, 1971, pp. 1, 12–15.
57. Joseph Wood Krutch, *The Measure of Man: On Freedom, Human Values, Survival and the Modern Temper* (New York: Grossett & Dunlap, 1954), p. 61.
58. BFS and Rogers, "Some Issues Concerning the Control of Human Behavior," p. 1063.
59. For an influential book that discusses the overwhelming strength of the liberators in American political history, see Lewis Hartz, *The Liberal Tradition in America: An Interpretation of American Political Thought Since the Revolution* (New York: Harcourt Brace, 1955). More recently, scholars have argued that a fundamental political dialogue about political philosophy and national destiny continued to the 1850s or even the 1890s. During the twentieth century, however, debate about core differences in political philosophy in the two-party system has dissolved. See Major Wilson, "The Concept of Time and Political Dialogue in the United States, 1828–1848," *American Quarterly* 19 (Winter 1967): 629–64; and Russell L. Hanson, *The Democratic Imagination in America: Conversations with Our Past* (Princeton, N.J.: Princeton University Press, 1985).
60. The historian Perry Miller was one of the first modern scholars to recognize the central place of a community ethic in Puritan thought. See his *The New*

England Mind: The Seventeenth Century (Boston: Beacon Press, 1954), esp. the chapter entitled "The Social Covenant," pp. 398–431. Other treatments of this theme include Sacvan Bercovitch, *The American Jeremiad* (Madison: University of Wisconsin Press, 1978); Loren Baritz, *City on a Hill: A History of Ideas and Myths in America* (New York: Wiley, 1964); Gordon Wood, *The Creation of the American Republic, 1776–1787* (Chapel Hill: University of North Carolina Press, 1969); and R. Jackson Wilson, *In Quest of Community: Social Philosophy in the United States, 1860–1920* (New York: Wiley, 1968).

61. B. F. Skinner, *Science and Human Behavior* (New York: Macmillan, 1953), p. 438.
62. BFS, *Beyond Freedom and Dignity,* p. 201.
63. BFS and Rogers, "Some Issues Concerning the Control of Human Behavior," p. 1061.
64. BFS, *Science and Human Behavior,* pp. 445–46.
65. BFS, *Beyond Freedom and Dignity,* p. 160.
66. B. F. Skinner, "Freedom and Control of Men," *American Scholar* 25 (1956): 47–65.
67. The difficulty Americans have had in establishing social and intellectual balance between freedom and control is a crucial interpretive problem in American history. Several outstanding works address this topic: Miller, *The New England Mind: From Colony to Province* (Boston: Beacon Press, 1961); Kenneth Lockridge, *A New England Town: The First One Hundred Years* (New York: Norton, 1970); Wood, *The Creation of the American Republic, 1776–1787*; Robert E. Shalhope, *The Roots of American Democracy: American Thought and Culture, 1760–1800* (Boston: Twayne, 1990); George Fredrickson, *The Inner Civil War: American Intellectuals and the Crisis of Union* (New York: Harper, 1965); Wilson, *In Quest of Community*; Robert Wiebe, *The Search for Order 1877–1920* (New York: Hill & Wang, 1967); Robert Wiebe, *The Segmented Society: An Introduction to the Meaning of America* (New York: Oxford University Press, 1975); and Ketcham, *Individualism and Public Life.*
68. Don Browning, "Pro Controls," *Christian Century* 88 (Sept. 22, 1971): 1116.

CHAPTER 10

1. George D. Wright, "A Further Note on Ranking the Important Psychologists," *American Psychologist* 25 (July 1970): pp. 650–51. In a second list, "Contemporary Contributors," Skinner was ranked first.
2. Donald A. Cook, "B. F. Skinner, the Man with Pigeons and Persistence," *Bostonia* 1 (Jan./Feb. 1991): 58.
3. Rose Kushner to Gay [?—illegible], Aug. 5, 1971, HA.
4. Skinner planned the fight with Keller as a "skit" for a banquet at the Midwestern Association of Behavioral Analysis (MABA) in early 1978. See BFS to FSK, Jan. 9, 1978, BA.
5. BFS to Arthur H. Brayfield, Apr. 18, 1968, HA.
6. Interview with BFS, Mar. 10, 1990. In an interview on Dec. 19, 1988, Skinner

addressed the charge that he was not particularly supportive of his graduate students. He felt that the "greatest compliment" he could give them was to allow them to "work on their own. . . . It was up to the student to establish himself or herself as a thinker."

7. An analysis of the reasoning behind the establishment of TIBA was recently written and shared with me by co-author Lawrence E. Fraley. See Stephen F. Ledoux and Lawrence E. Fraley, "The Emergence of the Discipline of Behaviorology as a Comprehensive Natural Science for Cultural Evolution: A Behaviorological History and Manifesto," to be published in TIBA's new journal, *Behaviorology.* Skinner was not especially enthusiastic about the establishment of this new organization, feeling that behaviorists should still struggle to make their case within Division 25 and ABA. As much as he disliked developments in those organizations, he disliked breaking altogether from mainstream organizations because of a negative effect on job opportunities for behaviorists.
8. See Julie S. Vargas, "B. F. Skinner: Father, Grandfather, Behavior Modifier," *Human Behavior* (Jan./Feb. 1972): 16–23. Skinner had dedicated *Beyond Freedom and Dignity* to his granddaughter—for "Justine and Her World."
9. Interview with BFS, Mar. 8, 1990. (Julie Vargas also attended and occasionally participated.)
10. Ibid.
11. BFS to John Hutchens, Aug. 12, 1975, HA. Skinner sent a note of thanks for the book that read:

Like pigeon and rat I was all box'd and wall'd in;
Control of my world was a good deal too thorough.
And so when a number of old friends were called in,
They broadened that world from a box to a Walden
Within a Walden, *in box, first edition by Thoreau. (Card, n.d., BA)*

12. Transcript of Yorkshire Film, 1977, HA. Became BBC film broadcast spring 1977 in England.
13. BFS to Leslie Reid, Aug. 9, 1972, HA.
14. Interview with BFS, Dec. 12, 1989.
15. Interview with BFS, Dec. 14, 1989.
16. BFS, "Wagner," Oct. 20, 1981, BA.
17. Interview with BFS, Aug. 13, 1990.
18. Interview with BFS, Dec. 14, 1989.
19. From the paper on his autobiography delivered in August 1979 at the APA, n.d., HA.
20. BFS, *Matter of Consequences,* p. 411.
21. Transcript of Yorkshire film, 1977, HA.
22. BFS to Silvia Saunders, Apr. 5, 1976, HA.
23. Silvia Saunders to BFS, Mar. 28, 1976, HA.

24. BFS, *About Behaviorism* (New York: Knopf, 1974), pp. 4–5.
25. See Robert Kirsch, *Los Angeles Times* [?] 1974, HA.
26. See Gail Boyer, *St. Louis Post-Dispatch* [?] 1974, HA.
27. BFS to Donald Bullock, Apr. 13, 1973, HA.
28. The occasion at Reed was a symposium on "The Control of Behavior: Legal, Scientific and Moral Dilemmas" in March 1975; the paper was later published in the *Criminal Law Bulletin* 11 (1975): 623–36. See also BFS, "The Holy Grail," Apr. 1, 1972, BA.
29. The paper was delivered in June 1973 and published in the *Thoreau Society Bulletin* 122 (Winter 1973): 1–3.
30. BFS, *Reflections on Behaviorism and Society* (Englewood Cliffs, N.J.: Prentice-Hall, 1978); *Upon Further Reflection* (Englewood Cliffs, N.J.: 1987); and *Recent Issues in the Analysis of Behavior* (Columbus, Ohio: Merrill, 1989).
31. Interview with BFS, Mar. 9, 1990.
32. See Charles Catania and Stevan Harnad, eds., *The Selection of Behavior, the Operant Behaviorism of B. F. Skinner: Comments and Consequences* (Cambridge: Cambridge University Press, 1989). The six "Canonical Papers" were "Selection by Consequences," "Methods and Theories in the Experimental Analysis of Behavior," "The Operational Analysis of Psychological Terms," "An Operant Analysis of Problem Solving," "Behaviorism at Fifty," and "The Phylogeny and Ontogeny of Behavior." These papers, as well as Skinner's responses to critics, confirm that the late Skinner defended his science as one that had never neglected heredity, but he firmly maintained that there was no inner agent that shaped a behavior once it appeared. Behavior was selected by environmental effects on the organism. This book affirmed that if "humanist intellectuals" reject Skinner's social argument in *Beyond Freedom and Dignity,* they are still very much interested in the philosophical and scientific implications of his work, both theoretical and experimental.
33. Interview with BFS, July 12, 1988. Skinner sent early fragments of his ethics book to several interested people. He gave up on it in early 1990 because he believed he had said much of the same thing elsewhere.
34. BFS and Margaret E. Vaughn, *Enjoy Old Age* (New York: Norton, 1983).
35. BFS, "The Holy Grail."
36. In a chapter entitled "Intellectual Self-Management in Old Age," Skinner quoted the Latin American novelist Jorge Luis Borges's lament, " 'What can I do at 71 except plagiarize myself?' " He insisted, however, "one *can* say something new. Creative verbal behavior is not produced by exercising creativity; it is produced by skillful self-management. . . . You are less likely to plagiarize yourself if you move into a new field or a new style" (*Upon Further Reflection,* p. 153). The chapter first appeared in the *American Psychologist* 38 (Mar. 1983): 239–44.
37. Interview with Richard Herrnstein, Aug. 14, 1990.
38. BFS to Robert Randall, Mar. 3, 1977, HA; and Robert Randall to BFS, Mar. 28, 1977, HA.

39. BFS to Richard H. C. Seabrook, Jan. 21, 1976, HA.
40. BFS, "Death," July 4, 1981, BA.
41. BFS, "Brain," Apr. 10, 1988, BA.
42. Interview with BFS, Dec. 12, 1989.
43. Interview with BFS, Aug. 13, 1990.
44. Interview with BFS, Mar. 7, 1990.
45. BFS, "Exposition," Feb. 7, 1974, BA.
46. For an excellent discussion of early scientific psychology, see Thomas Hardy Leahey, *A History of Psychology: Main Currents in Psychological Thought* (Englewood Cliffs, N.J.: Prentice-Hall, 1987), esp. "The Psychology of Consciousness," 179–205.
47. For more on the early behaviorists, see Leahey "The Rise of Behaviorism," in *A History of Psychology,* pp. 257–99.
48. BFS, "Dimensions of Mind," June 5, 1975, BA. For further discussion on his case against cognitive psychology, see "Why I Am Not a Cognitive Psychologist," in *Reflections on Behaviorism,* pp. 97–112; and "Cognitive Science and Behaviorism," in *Upon Further Reflection,* pp. 93–111.
49. BFS, "Myth," Oct. 31, 1976, BA.
50. BFS, untitled, Nov. 21, 1977, BA (his emphasis).
51. BFS to Susan M. Markle, Oct. 28, 1975, HA.
52. Interview with BFS for the *Boston Globe,* June 6, 1976, HA.
53. See, for example, "What Is Wrong with Daily Life in the Western World?" in *Upon Further Reflection,* pp. 15–31.
54. BFS, *Science and Human Behavior,* p. 4.
55. BFS, *Beyond Freedom and Dignity,* p. 1.
56. Ibid., p. 3.
57. Ibid., pp. 3–4.
58. FSK to BFS, Sept. 25, 1969, BA.
59. BFS to FSK, Sept. 30, 1969, BA.
60. BFS, *Reflections on Behaviorism,* p. 17. "Are We Free to Have a Future?" was first delivered to the Walgreen Conference on Education and Human Understanding at the University of Michigan in April 1973.
61. Ibid., p. 32.
62. BFS to David Palmer, Mar. 22, 1978, HA.
63. Interview with BFS, July 12, 1988.
64. BFS, *Recent Issues,* p. 118.
65. Interview with BFS, Mar. 9, 1990.
66. BFS, unpublished abstract of "Human Behavior and Democracy," Sept. [?], 1976, HA. Skinner delivered a paper with the same title to the APA in 1976, and it later appeared in *Psychology Today* (Sept. 1977) and was reprinted as the first chapter in *Reflections on Behaviorism,* pp. 3–15.
67. Three insightful works develop these themes in the formative period of American history. See Sacvan Bercovitch, *The American Jeremiad* (Madison: University of Wisconsin Press, 1978); Nathan O. Hatch, *The Sacred Cause of Liberty:*

Republican Thought and the Millennium in Revolutionary New England (New Haven, Conn.: Yale University Press, 1977); and Perry Miller, *Errand into the Wilderness* (Cambridge: Harvard University Press, 1956).

68. BFS to Phillip K. Bennett, Aug. 15, 1972, HA.
69. BFS to John R. Marquis, Dec. 6, 1971, HA.
70. BFS to R. H. Ankeny, Feb. 3, 1973, HA.
71. BFS to John B. Scott, Mar. 25, 1975, HA.
72. Tom Fitzpatrick, interview of BFS, *Chicago Sun Tribune,* Mar. 29, 1972, p. 24, HA.

BIBLIOGRAPHY

I have included works especially useful to my biographical reconstruction. For the most complete bibliography of works by Skinner, see Robert Epstein, "A Listing of the Published Works of B. F. Skinner, with Notes and Comments," *Behaviorism* 5 (1977): 99–110.

ALDRIDGE, JOHN W. "Afterthoughts on the 20's," *Commentary* (Nov. 1973): 40.

ALLEN, FREDERIC J., CHARLES ALLEN, AND CAROLYN F. ULRICH. *The Little Magazines: A History and a Bibliography*. Princeton, N.J.: Princeton University Press, 1946.

BAIDA, PETER. *Poor Richard's Legacy: American Business Values from Benjamin Franklin to Donald Trump*. New York: Morrow, 1990.

BARITZ, LOREN. *City on a Hill: A History of Ideas and Myths in America*. New York: Wiley, 1964.

———. "The Culture of the Twenties," in *The Development of an American Culture*, ed. Stanley Coben and Lorman Ratner, pp. 150–78. New York: St. Martin's Press, 1983.

BARNOUW, ERIK. *Tube of Plenty: The Evolution of American Television*. New York: Oxford University Press, 1982.

BENIGER, JAMES. *The Control Revolution: Technological and Economic Origins of the Information Society*. Cambridge, Mass.: Harvard University Press, 1986.

BERCOVITCH, SACVAN. *The Puritan Origins of the American Self*. New Haven, Conn.: Yale University Press, 1977.

———. *The American Jeremiad*. Madison: University of Wisconsin Press, 1978.

BERGMAN, LAWRENCE. *James Agee: A Life*. New York: Penguin Books, 1984.

BERMAN, LEWIS. *The Religion Called Behaviorism*. New York: Boni & Liveright, 1927.

BJORK, DANIEL W. *The Compromised Scientist: William James in the Development of American Psychology*. New York: Columbia University Press, 1983.

BLEDSTEIN, BURTON. *The Culture of Professionalism: The Middle Class and the Development of Higher Education in America*. New York: Norton, 1967.

BOAKES, ROBERT A. *From Darwin to Behaviourism: Psychology and the Minds of Animals*. Cambridge: Cambridge University Press, 1984.

BORING, E. G. "Edward Bradford Titchener," *American Journal of Psychology* 38 (1927): 489–506.

———. "Masters and Pupils among the American Psychologists," *American Journal of Psychology* 61 (1948): 527–34.

———. *A History of Experimental Psychology.* Englewood Cliffs, N.J.: Prentice-Hall, 1950.

———. *Psychologist at Large.* New York: Basic Books, 1961.

BORING, E. G., AND G. LINDZEY, EDS. *A History of Psychology in Autobiography,* vol. 5. New York: Appleton-Century-Crofts, 1967.

BRIDGMAN, PERCY. *The Logic of Modern Physics.* New York: Macmillan, 1927.

BROOKS, VAN WYCK. *The Flowering of New England 1815–1865.* New York: Dutton, 1937.

BROWNING, DON. "Pro Controls." *Christian Century* 88 (Sept. 22, 1971): 1116.

BRUCE, ROBERT V. *The Launching of American Science, 1846–1876.* New York: Knopf, 1987.

BUCKLEY, KERRY W. *Mechanical Man and the Beginnings of Behaviorism.* New York: Guilford Press, 1989.

BURTT, E. A. *The Metaphysical Foundations of Modern Physical Science: A Historical and Critical Essay.* London: Paul, Trench & Trubner, 1925.

CAPSHEW, JAMES H., AND ELIOT HEARST. "Psychology at Indiana University: From Bryan to Skinner." *Psychological Record* 30 (1980): 319–42.

CATANIA, CHARLES A. "*The Behavior of Organisms* as Work in Progress." *Journal of the Experimental Analysis of Behavior* 50 (1988): 277–81.

CATANIA, CHARLES A., AND STEVAN HARNAD. *The Selection of Behavior: The Operant Behaviorism of B. F. Skinner: Comments and Consequences.* Cambridge: Cambridge University Press, 1988.

CHOMSKY, NOAM. "A Review of B. F. Skinner's *Verbal Behavior.*" In *Readings in the Psychology of Language,* ed. L. A. Jakobovitz and M. S. Miron, pp. 142–71. New York: Prentice-Hall, 1967.

COHEN, I. BERNARD. *Benjamin Franklin: Scientist and Statesman.* New York: Scribner's, 1975.

COLEMAN, STEPHEN R. "Historical Context and Systematic Functions of the Concept of the Operant." *Behaviorism* 9 (1981): 207–26.

———. "B. F. Skinner: Systematic Iconoclast." *Gamut* 6 (Spring/Summer 1982): 53–75.

———. "B. F. Skinner, 1926–1928: From Literature to Psychology." *Behavior Analyst* 8 (1985): 77–92.

———. "When Historians Disagree: B. F. Skinner and E. G. Boring, 1930." *Psychological Record* 35 (1985): 301–14.

———. "Quantitative Order in B. F. Skinner's Early Research Program, 1928–1931." *Behavior Analyst* 10 (1987): 47–65.

CONANT, JAMES B. *The American High School Today: A First Report to Interested Citizens.* New York: McGraw-Hill, 1959.

CONOT, ROBERT. *Thomas A. Edison: A Streak of Luck.* New York: De Capo Press, 1979.

COOK, DONALD A. "B. F. Skinner: The Man with Pigeons and Persistence." *Bostonia* (Jan./Feb. 1991): 56–58.

———. "Can Computers Help?" *CBT Directions* 5 (July/Aug. 1992): 9–13.

COWLEY, MALCOLM. *Exile's Return.* New York: Viking, 1951.

CREMIN, LAWRENCE. *The Transformation of American Education: Progressivism in American Education, 1876–1957.* New York: Knopf, 1961.

CROSS, WHITNEY R. *The Burned-Over District: The Social and Intellectual History of Enthusiastic Religion in Western New York, 1800–1850.* New York: Harper & Row, 1965.

DEGLER, CARL N. *At Odds: Women and the Family in America from the Revolution to the Present.* Oxford: Oxford University Press, 1980.

DEWS, PETER D., ED. *Festschrift for B. F. Skinner.* New York: Appleton-Century-Crofts, 1970.

DINSMOOR, JAMES A. "A Visit to Bloomington: The First Conference on the Experimental Analysis of Behavior." *Journal of the Experimental Analysis of Behavior* 48 (Nov. 1987): 441–45.

———. "In the Beginning." *Journal of the Experimental Analysis of Behavior* 50 (Nov. 1988): 287–96.

———. "Keller and Schoenfeld's *Principles of Psychology,*" *Behavior Analyst* 12 (1989): 213–19.

———. "Setting the Record Straight: The Social Views of B. F. Skinner," *American Psychologist* 47 (Nov. 1992): 1454–63.

DONALDSON, SCOTT. *The Suburban Myth.* New York: Columbia University Press, 1969.

ELLSON, DOUGLAS G. "The Concept of the Reflex Reserve." *Psychological Review* 46 (Nov. 1939): 566–75.

ELMS, ALAN C. "Skinner's Dark Year and *Walden Two.*" *American Psychologist* 36 (1981): 470–79.

EPSTEIN, ROBERT. "A Listing of the Published Works of B. F. Skinner, with Notes and Comments." *Behaviorism* 5 (1977): 99–110.

———. *Notebooks of B. F. Skinner.* Englewood Cliffs, N.J.: Prentice-Hall, 1980.

ERIKSON, ERIK. *Young Man Luther: A Study in Psychoanalysis and History.* New York: Norton, 1958.

EVANS, RAND. "E. B. Titchener and His Lost System." *Journal of the History of Behavioral Sciences* 8 (1972): 168–80.

EVANS, RICHARD. *Dialogue with B. F. Skinner.* New York: Praeger, 1981.

FELLMAN, MICHAEL. *The Unbounded Frame: Freedom and Community in Nineteenth-Century American Utopianism.* Westport, Conn.: Greenwood Press, 1973.

FISHMAN, DANIEL H., FREDERICK ROTGERS, AND CYRIL M. FRANKS, EDS. *Paradigms of Behavior Therapy: Present and Promise.* New York: Springer, 1988.

FISHMAN, STEVE. "The Town B. F. Skinner Boxed." *In Health* 5 (Jan./Feb. 1991): 50–60.

FOX, DIXON RYAN. *The Decline of Aristocracy in the Politics of New York.* New York: Longmans, Green, 1919.

FRY, EDWARD B. *Teaching Machines and Programmed Instruction: An Introduction.* New York: McGraw-Hill, 1963.

FULLER, PAUL R. "Professors Kantor and Skinner—The 'Grand Alliance' of the 40's." *Psychological Record* 23 (1973): 318–24.

GALBRAITH, JOHN KENNETH. *The Affluent Society.* Boston: Houghton Mifflin, 1958.

GANS, HERBERT. *The Levittowners.* New York: Pantheon, 1967.

GARDNER, JUDITH KRIEGER. *Readings in Developmental Psychology.* New York: Little, Brown, 1978.

GIEDION, SIEGFRIED. *Mechanism Takes Command.* New York: Norton, 1969.

GILBERT, JAMES B. *Another Chance: Postwar America 1945–1985.* New York: Dorsey Press, 1986.

GOODELL, RAE. *The Visible Scientists.* New York: Little, Brown, 1975.

GREEN, JOHN D. "B. F. Skinner's Technology of Teaching." *Classroom Computer Learning* (Feb. 1984): 23–24, 28–29.

GRIFFEN, CLYDE. "The Progressive Ethos." In *The Development of an American Culture,* ed. Stanley Coben and Lorman Ratner, pp. 120–49. Englewood Cliffs, N.J.: Prentice-Hall, 1970.

GUTTMAN, NORMAN. "On Skinner and Hull: A Reminiscence and Projection." *American Psychologist* (May 1977): 321–28.

HALE, NATHAN, JR. *Freud and the Americans: The Beginnings of Psychoanalysis in the United States, 1876–1917.* New York: Oxford University Press, 1971.

HALL, ELIZABETH. "Will Success Spoil B. F. Skinner?" *Psychology Today* (Nov. 1972): 65–72, 130.

HANSON, RUSSELL L. *The Democratic Imagination in America: Conversations with Our Past.* Princeton, N.J.: Princeton University Press, 1985.

HARRIS, NEIL. *The Artist in American Society: The Formative Years, 1790–1860.* New York: Braziller, 1968.

HARTZ, LEWIS. *The Liberal Tradition in America: An Interpretation of American Political Thought since the Revolution.* New York: Harcourt Brace Jovanovich, 1955.

HATCH, NATHAN O. *The Sacred Cause of Liberty: Republican Thought and the Millennium in Revolutionary New England.* New Haven, Conn.: Yale University Press, 1977.

HEARST, ELIOT, AND JAMES H. CAPSHEW, EDS. *Psychology at Indiana University: A Centennial Review and Compendium.* Bloomington: Indiana University Department of Psychology, 1988.

HERRNSTEIN, RICHARD. "Reminiscences Already." *Journal of the Experimental Analysis of Behavior* 48 (Nov. 1987): 448–53.

HIGARD, ERNEST R. "Review of *The Behavior of Organisms.*" *Psychological Bulletin* 36 (1939): 121–25.

HILFER, ANTHONY CHANNELL. *The Revolt from the Village, 1915–1930.* Chapel Hill, N.C.: University of North Carolina Press, 1969.

HINDLE, BROOK. *Emulation and Invention.* New York: New York University Press, 1982.

HOFFMAN, FREDERIC J. *The Twenties: American Writing in the Postwar Decade.* New York: Free Press, 1965.

HOLLINGER, DAVID A. "The Knower and the Artificer." In *Modernist Culture and American Life,* ed. Joseph Singal, pp. 42–69. Belmont, Calif.: Wadsworth, 1991.

HOWARD, GERALD, ED. *The Sixties: Art, Politics and Media in Our Most Explosive Decade.* New York: Paragon House, 1991.

HUGHES, THOMAS P. *American Genesis: A Century of Invention and Technological Enthusiasm, 1870–1970.* New York: Viking, 1989.

HUTCHENS, JOHN K. "Way Back Then, and Later." *Hamilton Alumni Review* 42 (Dec. 1976): 12–18.

JACKSON, KENNETH. *Crabgrass Frontier: The Suburbanization of America.* New York: Oxford University Press, 1985.

JONES, LANDON. *Great Expectations: America and the Baby Boom Generation.* New York: Ballentine, 1986.

KAPLAN, HAROLD. "Life in a Cage." *Commentary* 53 (Feb. 1972): 82–84, 86.

KELLER, FRED S. *Summers and Sabbaticals: Selected Papers on Psychology and Education.* Champaign, Ill.: Research Press, 1977.

———. "A Fire in Schermerhorn Extension." *Behavior Analyst* 9 (1986): 139–46.

———. "Burrhus Frederic Skinner (1904–1990) (A Thank You)." *Journal of the Experimental Analysis of Behavior* 54 (1990): 155–58.

KEEHN, J. D., ED. *Walden Three.* Downsview, Ontario: Master Press, 1984.

KENNER, HUGH. *A Homemade World: The American Modernist Writers.* New York: Knopf, 1975.

KETCHAM, RALPH. *Individualism and Public Life: An American Dilemma.* New York: Blackwell, 1987.

KIDD, RONALD V., AND NATALICIO, LUIZ. "An Interbehavioral Approach to Operant Analysis." *Psychological Record* 32 (1982): 41–59.

KITCHNER, RICHARD F. "Are There Molar Psychological Laws?" *Philosophy of the Social Sciences* 6 (1976): 143–56.

KLAW, SPENCER. "Harvard's Skinner: The Last of the Uptopians." *Harper's,* Apr. 1963, pp. 45–51.

KNAPP, TERRY J. "Behaviorism in a Hundred Years of Psychology: A Review of a Century of Psychology as a Science." *Behaviorism* 15 (1987): 171–74.

KOLKO, GABRIEL. *The Triumph of Conservatism: A Re-interpretation of American History, 1900–1916.* New York: Free Press, 1963.

KONORSKI, J., AND S. MILLER. "On Two Types of Conditioned Reflex." *Journal of General Psychology* 12 (1935): 66–77.

KOONRADT, PAUL. *Dance Out the Answer.* New York: Longmans, Green, 1932.

KRANTZ, DAVID L. "Schools and Systems: The Mutual Isolation of Operant and Non-operant Psychology as a Case Study." *Journal of the History of the Behavioral Sciences* 8 (1972): 86–102.

KRASNER, LEONARD. "The Future and the Past in the Behaviorism-Humanism Dialogue." *American Psychologist* 33 (1978): 799–804.

———. "Paradigm Lost: On a Historical/Sociological/Economic Perspective." In

Paradigms in Behavior Therapy: Present and Promise, ed. Daniel B. Fishman, Frederick Rotgers, Cyril M. Franks, pp. 23–44. New York: Springer, 1988.

KRUTCH, JOSEPH W. *The Modern Temper: A Study and a Confession.* New York: Harcourt Brace, 1929.

———. *The Measure of Man: On Freedom, Human Values, Survival and the Modern Temper.* New York: Grossett & Dunlap, 1954.

KUHN, THOMAS S. *The Structure of Scientific Revolutions.* Chicago: Chicago University Press, 1962.

LATIES, VICTOR G. "Society for the Experimental Analysis of Behavior: The First Thirty Years (1957–1987)." *Journal of the Experimental Analysis of Behavior* 48 (Nov. 1987): 495–512.

LEAHEY, THOMAS H. *A History of Psychology: Main Currents in Psychological Thought.* Englewood Cliffs, N.J.: Prentice-Hall, 1987.

LEIBER, JUSTIN. *Noam Chomsky: A Philosophic Overview.* Boston: Twayne, 1975.

LEVINSON, DANIEL J. *The Seasons of a Man's Life.* New York: Ballantine Books, 1978.

LEWIS, R. W. B. *The American Adam: Innocence, Tragedy and Tradition in the Nineteenth Century.* Chicago: University of Chicago Press, 1964.

LEWIS, SINCLAIR. *Main Street.* New York: Harcourt Brace, 1920.

LINK, ARTHUR S. "What Happened to the Progressive Movement in the 1920s?" *American Historical Review* 64 (July 1959): 833–51.

LOCKRIDGE, KENNETH. *A New England Town: The First One Hundred Years.* New York: Norton, 1970.

LUMSDAINE, A. A., AND ROBERT GLASER. *Teaching Machines and Programmed Instruction: A Source Book.* Washington, D.C.: National Education Association, 1960.

MACCORQUODALE, KENNETH. "B. F. Skinner's *Verbal Behavior:* A Retrospective Appreciation." *Journal of the Experimental Analysis of Behavior* 12 (1969): 831–41.

———. "On Chomsky's Review of Skinner's *Verbal Behavior.*" *Journal of the Experimental Analysis of Behavior* 13 (1970): 83–99.

MACHAN, TIBOR R. *The Pseudo-Psychology of B. F. Skinner.* New Rochelle, N.Y.: Arlington House, 1974.

MACHEN, ARTHUR. *A Hill of Dreams.* London: Baker, 1968.

MACKENZIE, BRIAN D. *Behaviorism and the Limits of Scientific Method.* Atlantic Highlands, N.J.: Humanities Press, 1977.

MCPHERSON, JAMES M. *Battle Cry of Freedom: The Civil War Era.* New York: Oxford University Press, 1988.

MANCHESTER, WILLIAM. *The Glory and the Dream: A Narrative History of America, 1932–1972.* Boston: Little, Brown, 1974.

MARSHALL, JOHN C. "Language Learning, Language Acquisition, or Language Growth?" In *Noam Chomsky: Consensus and Controversy,* ed. Sohan Modgil and Celia Modgil, pp. 41–49. New York: Falmer Press, 1970.

MATTHEISSEN, F. O. *American Renaissance: Art and Expression in the Age of Emerson and Whitman.* New York: Oxford University Press, 1941.

MEDAWAR, PETER. *The Limits of Science.* New York: Oxford University Press, 1984.

Meyer, Donald. *The Positive Thinkers: Religion as Pop Psychology from Mary Baker Eddy to Oral Roberts.* New York: Pantheon, 1980.

Michael, Jack. "Verbal Behavior." *Journal of the Experimental Analysis of Behavior* 42 (1984): 363–76.

Miller, Perry. *The New England Mind: From Colony to Province.* Cambridge, Mass.: Harvard University Press, 1953.

———. *Errand into the Wilderness.* Cambridge, Mass.: Harvard University Press, 1956.

Mintz, Steven, and Susan Kellogg. *Domestic Revolutions: A Social History of American Family Life.* New York: Free Press, 1988.

Newman, Bobby. *The Reluctant Alliance: Behaviorism and Humanism.* Buffalo, N.Y.: Prometheus Books, 1992.

North, Douglas C. *The Economic Growth of the United States, 1790–1860.* New York: Norton, 1966.

Nye, Robert D. *Three Psychologies: Perspectives from Freud, Skinner, and Rogers.* Monterey, Calif.: Brooks/Cole, 1981.

———. *The Legacy of B. F. Skinner: Concepts and Perspectives, Controversies and Misunderstandings.* Pacific Grove, Calif.: Brooks/Cole, 1992.

O'Donnell, John M. *The Origins of Behaviorism: American Psychology, 1870–1920.* New York: New York University Press, 1985.

Ogden, C. K., and I. A. Richards. *The Meaning of Meaning.* New York: Harcourt Brace, 1923.

Pauly, Philip J. *Controlling Life: Jacques Loeb and the Engineering Ideal in Biology.* Berkeley: University of California Press, 1990.

Pavlov, Ivan P. *Conditioned Reflexes.* Trans. G. V. Anrep. New York: Oxford University Press, 1927.

Peck, James, ed. *The Chomsky Reader.* New York: Pantheon Books, 1987.

Pells, Richard. *Radical Visions and American Dreams: Culture and Social Thought during the Depression Years.* New York: Wesleyan University Press, 1973.

Pessen, Edward. *Jacksonian America: Society, Personality and Politics.* Homewood, Ill.: Dorsey Press, 1978.

Pilkington, Walter. *Hamilton College, 1812–1962.* Deposit, N.Y.: Courier Printing, 1962.

Pomeroy, Elsie M. *William Saunders and His Five Sons: The Story of the Marquis Wheat Family.* Toronto: Ryerson Press, 1956.

Pressey, Sidney. "A Simple Device for Teaching, Testing and Research." *School and Society* 23 (1926): 373–76.

Pritchard, William H. *Frost: A Literary Life Reconsidered.* New York: Oxford University Press, 1984.

Puligandla, H. *Fact and Fiction in B. F. Skinner's* Science and Utopia. St. Louis, Mo.: Green, 1974.

Rand, Ayn. *Philosophy: Who Needs It?* New York: Signet, 1984.

Rescher, Nicholas. *Unpopular Essays on Technological Progress.* Pittsburgh, Penn.: University of Pittsburgh Press, 1980.

RICE, BERKELEY. "Skinner Agrees He Is the Most Important Influence in Psychology." *New York Times Magazine* (Mar. 17, 1968): 27.

ROBINSON, DANIEL N. *An Intellectual History of Psychology.* Madison: University of Wisconsin Press, 1986.

RUPPERT, PAUL. *B. F. Skinner's Walden Two: A Critical Commentary.* New York: Simon & Schuster, 1976.

RUSSELL, BERTRAND. *Philosophy.* New York: Norton, 1927.

SAGAL, PAUL T. *Skinner's Philosophy.* Washington, D.C.: University Press of America, 1981.

SCHAEFFER, FRANCIS A. *Back to Freedom and Dignity.* Downers Grove, Ill.: Inter-Varsity Press, 1972.

SCHOENFELD, WILLIAM N. "Reminiscences You Say." *Journal of the Experimental Analysis of Behavior* 48 (Nov. 1978): 464–68.

SCHWARTZ, BARRY. *Behaviorism, Science, and Human Nature.* New York: Norton, 1982.

SCHWARTZ, BARRY, RICHARD SCHULDENFREI, AND HUGH LACEY. "Operant Psychology as Factory Psychology." *Behaviorism* 6 (1978): 229–54.

SCRIVEN, MICHAEL. "A Study in Radical Behaviorism." In *Minnesota Studies in the Philosophy of Science,* vol. 1, ed. Herbert Feigl and Michael Scriven, pp. 88–130. Minneapolis: University of Minnesota Press, 1956.

SEGAL, HOWARD P. *Technological Utopianism in American Culture.* Chicago: University of Chicago Press, 1985.

SENNETT, RICHARD. "Review of *Beyond Freedom and Dignity.*" *New York Times Book Review,* Oct. 17, 1971, pp. 1, 12–15.

SHALHOPE, ROBERT E. *The Roots of American Democracy: American Thought and Culture, 1760–1800.* Boston: Twayne, 1990.

SHEA, D. B., JR. "B. F. Skinner: The Puritan Within." *Virginia Quarterly Review* 50 (1974): 416–37.

SINGAL, DANIEL JOSEPH, ED. *Modernist Culture and American Life.* Belmont, Calif.: Wadsworth, 1991.

SKINNER, B. F. "The Concept of the Reflex in the Description of Behavior." *Journal of General Psychology* 5 (1931): 427–58.

———. "Drive and Reflex Strength." *Journal of General Psychology* 6 (1932): 22–48.

———. "On the Rate of Formation of a Conditioned Reflex." *Journal of General Psychology* 8 (1932): 274–86.

———. "On the Rate of Extinction of a Conditioned Reflex." *Journal of General Psychology* 9 (1933): 114–29.

———. "The Measurement of Spontaneous Activity." *Journal of General Psychology* 9 (1933): 3–23.

———. "The Rate of Establishment of a Discrimination." *Journal of General Psychology* 9 (1933): 302–50.

———. "Has Gertrude Stein a Secret?" *Atlantic Monthly* 153 (1934): 50–57.

———. "The Generic Nature of the Concepts of Stimulus and Response." *Journal of General Psychology* 12 (1935): 40–65.

———. "Two Types of Conditioned Reflexes and a Pseudo Type." *Journal of General Psychology* 12 (1935): 66–77.
———. "The Verbal Summator and a Method for the Study of Latent Speech." *Journal of Psychology* 2 (1936): 71–107.
———. "Two Types of Conditioned Reflex: A Reply to Konorski and Miller." *Journal of General Psychology* 16 (1937): 272–79.
———. *The Behavior of Organisms: An Experimental Analysis.* New York: Appleton-Century-Crofts, 1938.
———. "The Alliteration in Shakespeare's Sonnets: A Study of Literary Behavior." *Psychological Record* 3 (1939): 186–92.
———. "Hull's *Principles of Behavior.*" *American Journal of Psychology* 57 (1944): 276–81.
———. "Baby in a Box." *Ladies' Home Journal,* Oct. 1945, pp. 30–31, 135–36, 138.
———. *Walden Two.* New York: Macmillan, 1948.
———. "Are Learning Theories Necessary?" *Psychological Review* 57 (1950): 193–216.
———. *Science and Human Behavior.* New York: Macmillan, 1953.
———. "The Science of Learning and the Art of Teaching." *Harvard Educational Review* 24 (1954): 86–97.
———. "Freedom and the Control of Men." *American Scholar* (Winter 1955–1956): 47–65.
———. "A Case Study in Scientific Method." *American Psychologist* 11 (1956): 221–33.
———. *Verbal Behavior.* New York: Appleton-Century-Crofts, 1957.
———. "Teaching Machines." *Science* 128 (1958): 969–77.
———. *Cumulative Record.* New York: Appleton-Century-Crofts, 1959.
———. "John Broadus Watson, Behaviorist." *Science* 129 (1959): 197–98.
———. "The Flight from the Laboratory." In B. F. Skinner, *Cumulative Record.* New York: Appleton-Century-Crofts, 1959.
———. "Pigeons in a Pelican." *American Psychologist* 15 (1960): 28–37.
———. "The Design of Cultures." *Daedalus* 90 (1961): 534–46.
———. "Behaviorism at Fifty." *Science* 140 ((1963): 951–58.
———. "Operant Behavior." *American Psychologist* 18 (1963): 503–15.
———. "Reflections on a Decade of Teaching Machines." *Teachers College Record* 65 (1963): 168–77.
———. "The Phylogeny and Ontogeny of Behavior." *Science* 153 (1966): 1205–13.
———. "What Is the Experimental Analysis of Behavior?" *Journal of the Experimental Analysis of Behavior* 9 (1966): 213–18.
———. "Some Responses to the Stimulus 'Pavlov.'" *Conditioned Reflex* 1 (1966): 74–78.
———. "B. F. Skinner." In *A History of Psychology in Autobiography,* vol. 5, ed. E. G. Boring and G. Lindzey, pp. 387–413. New York: Appleton-Century-Crofts, 1967.
———. "The Design of Experimental Communities." *International Encyclopedia of the Social Sciences,* vol. 16, pp. 271–75. New York: Macmillan, 1968.
———. *The Technology of Teaching.* New York: Appleton-Century-Crofts, 1968.

———. *The Contingencies of Reinforcement: A Theoretical Analysis.* New York: Appleton-Century-Crofts, 1969.

———. "Creating the Creative Artist." In *On the Future of Art,* ed. A. J. Toynbee et al., pp. 61–75. New York: Viking, 1970.

———. *Beyond Freedom and Dignity.* New York: Knopf, 1971.

———. "Answers for My Critics." In *Beyond the Punitive Society,* ed. H. Wheeler, pp. 256–66. San Francisco: Freeman, 1973.

———. *About Behaviorism.* New York: Knopf, 1974.

———. *Particulars of My Life.* New York: Knopf, 1976.

———. "Walden Two Revisited." In *Walden Two.* New York: Macmillan, 1976.

———. "Why I Am Not a Cognitive Psychologist." *Behaviorism* 5 (1977): 1–10.

———. *Reflections on Behaviorism and Society.* Englewood Cliffs, N.J.: Prentice-Hall, 1978.

———. *The Shaping of a Behaviorist.* New York: Knopf, 1979.

———. "Origins of a Behaviorist." *Psychology Today*, Sept. 1983, pp. 22–33.

———. *A Matter of Consequences.* New York: Knopf, 1984.

———. *Upon Further Reflection.* Englewood Cliffs, N.J.: Prentice-Hall, 1987.

———. *Recent Issues in the Analysis of Behavior.* Columbus, Ohio: Merrill, 1989.

Skinner, B. F., and T. C. Barnes. "The Progressive Increase in Geotropic Response of the Ant *Aphaenogaster." Journal of General Psychology* 4 (1930): 102–12.

Skinner, B. F., and William T. Crozier. "Review of Franklin Fearing's *Reflex Action: A Study in the History of Physiological Psychology." Journal of General Psychology* 5 (1931): 125–29.

Skinner, B. F., and W. T. Heron. "Effects of Caffeine and Benzedrine upon Conditioning and Extinction." *Psychological Record* 1 (1937): 340–46.

———. "An Apparatus for the Study of Animal Behavior." *Psychological Record* 3 (1939): 166–76.

Skinner, B. F., with Charles Ferster. *Schedules of Reinforcement.* New York: Appleton-Century-Crofts, 1957.

Skinner, B. F., and James Holland. *The Analysis of Behavior.* New York: McGraw-Hill, 1961.

Skinner, B. F., and Carl Rogers. "Some Issues Concerning the Control of Human Behavior: A Symposium." *Science* 124 (1956): 1057–66.

Skinner, B. F., with Margaret E. Vaughan. *Enjoy Old Age.* New York: Knopf, 1983.

Smith, Laurence D. *Behaviorism and Logical Positivism: A Reassessment of the Alliance.* Stanford, Calif.: Stanford University Press, 1986.

———. "On Prediction and Control: B. F. Skinner and the Technological Ideal of Science." *American Psychologist* 47 (Feb. 1992): 216–23.

Smith, Wendell T., and J. William Moore, eds. *Programmed Learning: Theory and Research.* Princeton, N.J.: Van Nostrand, 1962.

Stevick, Daniel B. *B. F. Skinner's* Walden Two: *Introduction and Commentary.* New York: Seabury Press, 1968.

STOCKER, RHAMANTHUS M. *Centennial History of Susquehanna County, Pennsylvania.* Baltimore: Regional, 1974.

SUSMAN, WARREN I. *Culture as History: The Transformation of American Society in the Twentieth Century.* New York: Pantheon, 1984.

TAYLOR, GEORGE R. *The Transportation Revolution: 1815–1860.* New York: Rinehart, 1951.

THOMPSON, LAWRENCE. *Robert Frost: The Early Years, 1874–1915.* New York: Holt, Rinehart & Winston, 1966.

———. *Robert Frost: The Years of Triumph, 1915–1938.* New York: Holt, Rinehart & Winston, 1970.

———. *Robert Frost: The Later Years, 1938–1963.* New York: Holt, Rinehart & Winston, 1976.

———, ED. *Selected Letters of Robert Frost.* New York: Holt, Rinehart & Winston, 1964.

THOMPSON, TRAVIS. "Retrospective Review: Benedictus Behavioral Analysis: B. F. Skinner's Magnum Opus at Fifty." *Contemporary Psychology* 33 (1988): 397–402.

THORSON, T. W. "A Rhetorical Analysis of the Discourse of B. F. Skinner and His Principal Critics." Ph.D. diss., Northwestern University, 1973.

TOLMAN, EDWARD. "Psychology versus Immediate Experience." *Philosophy of Science* 2 (1935): 356–80.

TRATCHENBURG, ALAN. *The Incorporation of America: Culture and Society in the Gilded Age.* New York: Hill & Wang, 1982.

TYLER, ALICE FELT. *Freedom's Ferment: Phases of American Social History from the Revolution to the Outbreak of the Civil War.* New York: HarperCollins, 1962.

VAN DOREN, CARL. "The Revolt from the Village: 1920," *Nation,* October 12, 1921, pp. 307–412.

VARGAS, ERNEST A., AND JULIE S. VARGAS. "Programmed Instruction and Teaching Machines." In *Designs for Excellence in Education,* ed. Richard P. West and L. A. Hamerlynack, pp. 33–69. Longment, Calif.: Sopris West, 1992.

VARGAS, JULIE S. "B. F. Skinner: Father, Grandfather, Behavior Modifier." *Human Behavior* (Jan./Feb. 1972): 16–23.

———. "B. F. Skinner: Fact and Fiction." *International Behaviorology Association Newsletter* 2 (Winter 1990): 8–11.

WARE, CAROLINE F. *Greenwich Village, 1920–1930.* New York: Harper & Row, 1935.

WATSON, JOHN B. "Psychology as the Behaviorist Views It." *Psychological Review* 20 (1913): 158–77.

———. *Behavior: An Introduction to Comparative Psychology.* New York: Holt, 1914.

———. *Psychological Care of Infant and Child.* New York: Horton, 1928.

WEBB, MRS. ARTHUR. "History of Susquehanna, Pa." [1853–1953], *Susquehanna Cheers 100 Years,* Susquehanna Centennial, Aug. 16–22, 1953 (week of centennial celebration).

WEIGEL, JOHN A. *B. F. Skinner.* Boston: Hall, 1977.

WHEELER, H., ED. *Beyond the Punitive Society.* San Francisco: Freeman, 1973.

WIEBE, ROBERT. *The Search for Order, 1877–1920.* New York: Hill & Wang, 1967.

———. *The Segmented Society: An Introduction to the Meaning of America.* New York: Oxford University Press, 1975.

WIKLANDER, NILS. *From Laboratory to Utopia: An Inquiry into the Early Psychology and Social Philosophy of B. F. Skinner.* Götenborg University: Götenborg University Press, 1989.

WILLARD, M. J., AND ROBERT EPSTEIN. "Our Most Unforgettable Character." *Behavior Analyst* 3 (1980): 35–39.

WILSON, MAJOR. "The Concept of Time and Political Dialogue in the United States, 1828–1848." *American Quarterly* 19 (Winter 1967): 629–64.

WILSON, R. JACKSON. *The Quest for Community: Social Philosophy in the United States, 1860–1920.* New York: Wiley, 1968.

WOOD, GORDON. *The Creation of the American Republic, 1776–1787.* Chapel Hill, N.C.: University of North Carolina Press, 1969.

YOUNG, R. M. *Mind, Brain and Adaptation in the Nineteenth Century.* New York: Clarendon Press, 1970.

ZINSSER, WILLIAM. *Extraordinary Lives: The Art and Craft of American Biography.* New York: Houghton Mifflin, 1986.

ZURIFF, GERALD E. *Behaviorism: A Conceptual Reconstruction.* New York: Columbia University Press, 1985.

Index